Mysterious Lights and Crop Circles

LINDA MOULTON HOWE

Cover and Frontispiece Photograph

Triple Julia Set discovered July 29, 1996, at Windmill Hill, Wiltshire, England, near Avebury, the largest circle and avenue of standing sarsen stones in the British Isles, constructed around 3700 B.C. for spiritual rituals. This fractal formation of 194 circles spiraled for 1,000 feet in diameter and is discussed further in Chapter 2. Photograph © 1996 by Steve Alexander.

Other Books by Author

An Alien Harvest

Glimpses of Other Realities, Vol. I: Facts & Eyewitnesses

Glimpses of Other Realities, Vol. II: High Strangeness

Back Cover Photograph

"Basket" wheat formation discovered August 6, 1999, at Bishops Cannings, Wiltshire, England. Aerial videograph taken by Ulrich Kox and © by Ulrich Kox and Peter Sorensen.

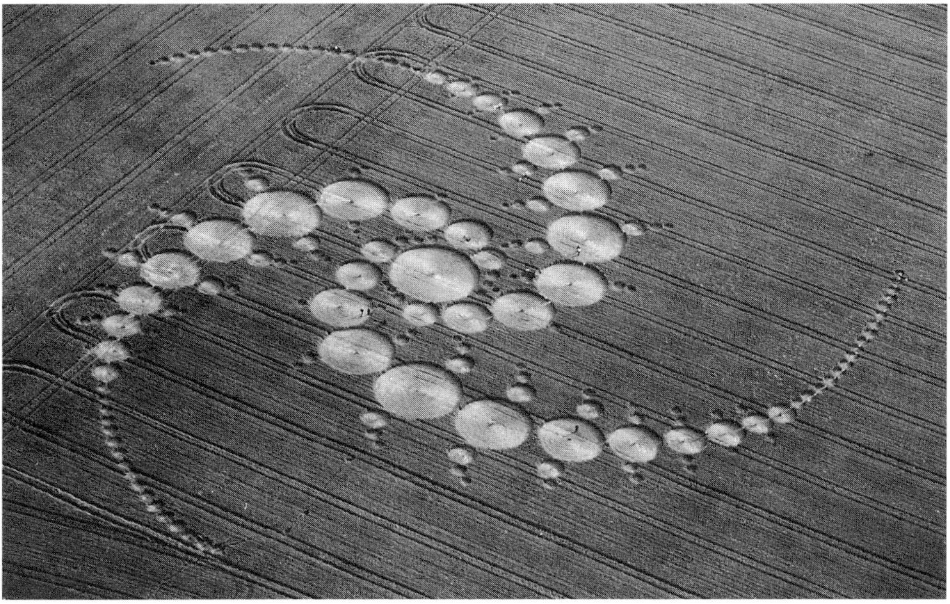

Whatever is doing these formations
is affecting the fundamental biochemistry
and biophysics of the plants.

—W. C. Levengood, Ph.D., biophysicist

First published 2000

First Edition, First Printing, 2000

Published in the United States of America at
Paper Chase Press
5721 Magazine Street, Ste. 152
New Orleans, Louisiana 70115

Book text is New Baskerville 11 point
with Dolphin titles.

Maps by Tim Bauer, Gettysburg, Pennsylvania.

Library of Congress Catalog Card Number: 00-133648

ISBN 1-879706-91-1

PAPER
CHASE
PRESS

Contents

Foreword

If you have bought, or even opened this excellent book, and find yourself reading its foreword, the chances are you do not have to be convinced that crop circles are mysterious, inexplicable and real. That is to say, they are not, for the most part, hoaxes — although the scientific community and the media seem committed to convincing the public that they are. Linda Moulton Howe wastes no time recapitulating the notorious Doug and Dave episode. She dismisses it in a single paragraph and gets on with exploring the complex, ongoing aftermath. And what she finds is a common thematic and symbolic thread, plus empirical evidence from scientific research that cannot be hoaxed.

If you did not know crop circles were glyphs incised into fields of grain and knew them only as photographs, you might think they were paintings produced by a genius abstract artist educated in some universal mystery religion or esoteric school. The figures speak directly to our human aesthetic/spiritual faculty, as does all art. The language spoken is at once familiar and strange. The crop circles resonate within, summoning up echoes of the philosophical, metaphysical and religious symbols used around the world since history began. Yet, rarely are they identical to what is known. Moreover, during the crop circle decades of the 1980s and 1990s, mysterious balls of light or light forms have been seen hovering about the formations, sometimes appearing on photographs and videotapes. Even lights and spheres invisible to the naked eye turn up on film and tape.

It's difficult to avoid the feeling that the circles and the circle makers, whoever or whatever they might be, are going to great lengths trying to tell us something. Possibly something important, and the inference is that the message must be positive since it is counterintuitive to associate so much beauty and harmony with negation. But what might that message be? And to what end? It is this positive aspect of the crop circle mystery that Linda Moulton Howe stresses while raising an important question about point and counterpoint. The symbol glyphs often have pentacles or geometries inside circles that suggest black and white magical rites. Could at least two players with different intents be competing on the chessboards of the fields? Thoroughly, thoughtfully, sensitively, she explores the innumerable resonances the circles set up with symbols we already know from the ancient traditions of Egypt, Mesopotamia, Islam, Christianity, Judaism, the Far East, and shamanic societies. Symbols specific to none, reminiscent of all. Remarkable! And then there are those lights! The book is gorgeously and richly illustrated.

As she weaves the symbols throughout the book with occasional near matches in ancient texts, a central theme emerges: that there is no positive force or phenomenon without its negative counterpart: matter/antimatter, active/passive, initiating/resisting. There is no Horus without Seth. That is a cosmological, psychological, physical and strictly practical truth. In ancient Egypt, Seth (Setekh) was Ego, principle of materialization and opposition, of negation, of matter forever at war with Spirit. Seth is the Dark Force that does not want to return to the Source. Seth is also a great god. After prodigious contentions with Horus, Seth is eventually conquered and *reconciled,* not killed or even banished. *Both* Seth and Horus bless and empower Pharaoh, the Divinity incarnated upon earth. An alchemical text captures the final stage of that process: "The fire goes out only when the work is complete and the whole tinctorial mass fuses with the glass and becomes luminous like the sun."

Without Seth in opposition, nothing happens or can happen. Seth is easily understood, so it is odd that everywhere he is so grievously misunderstood. When I turn on my computer in the morning, that blank screen is Seth, the Opposition, defying me to write words on it. We learn, all of us, mainly through negation. Praise is comforting. It's nice to be told that your work is admired. But the work is taken further largely through a recognition of the negative. If I do not know what's wrong, or what is inadequate, I cannot improve it, fix it, surmount it. Seth is always present — as he must be, should be — at least until the work is complete and that "tinctorial mass fuses with the glass and becomes luminous like the sun."

In the myth, Horus always conquers. In real life, it's not always like that. In fact, for roughly the last 4,000 years, the world has seen little of Horus, but a surfeit of Seth. In the myth, Seth takes many forms that have negative associations: scorpion, viper, crocodile, rampaging male hippo, jackass. Seth is also master of confusion, camouflage, deception. So it comes as no surprise when, in modern time, he dons Horus's own falcon mask of Resurrection and Return and poses as today's Messiah preaching his own gospel of Materialism and Meaninglessness. Then there's that church of glitzy barbarism he calls

"Progress" whose Jesuits are Science, Education and the Media. Its job: to defend the Unfaith.

Seen as a manifestation of the Sethian force, the otherwise puzzling opposition to crop circles and, indeed, to any of the subjects directly, tangentially or indirectly associated with crop circles becomes not only understandable, but inevitable. Unidentified flying objects, ESP in its many forms, "lost" civilizations, astrology, zero-point energy, homeopathy and other alternative therapies, reincarnation, higher consciousness and the doctrine of the Immortality of the Soul — all, in one way or another, threaten the authority of the Church of Progress and must be disavowed, discredited and disempowered.

And amid the mysterious crop circle messages to humanity that many sense come from positive intelligence, why has the Anti–Crop Circle campaign to conceal, obfuscate and discredit been so successful? They are so very beautiful, moving and seemingly harmless. Yet, they were trivialized when an uninformed public bought the Doug and Dave scenario wholesale. Strange! Dave even died in 1996. But back in 1991, those two elderly drinking buddies told a tabloid television reporter they had been using boards and ropes to produce crude patterns in two southern counties of England. The world press and media heaved a huge sigh of coordinated relief to have an easy answer for an awkward story that did not fit neatly into politically acceptable boxes. Enthusiastically, they publicized the hoaxers in headline banners and prime-time news. So that was it! No angelic forces, no UFOs, no earth spirits, not even whirlwinds. It was all Doug and Dave. Why, the logic went, if Doug and Dave could produce one hoax, then by extension, *all* the other several thousand crop circles around the world over the past two decades must also be hoaxes.

It hardly mattered that no hoaxer had ever come forward before to claim responsibility, raising great suspicions among more knowledgeable researchers that someone with vested interests in maintaining a status quo *without* crop formations had paid off the two gents to float a lie just like weather balloons and swamp gas. Journalists, whose job it is to check data and to look into all aspects of any given story before ratifying its veracity, were suddenly, as if by undivine decree, relieved of their customary responsibilities. What did it matter that thousands of hostile farmers in two dozen countries did not want their crops flattened, *period!* by whatever agency and had never, ever caught a hoaxer in the act? Nor did it matter that laboratory tests by highly credentialed scientists had found that the grain within the formations had undergone inexplicable biophysical and biochemical changes; nor that the complexity of new geometry theorems stunned professional mathematicians. Ah, yes! The mathematicians didn't know. But Doug and Dave knew. "Those studying the facts," writes Linda Moulton Howe, "knew the tabloid story was not true and were frustrated that the global media would so easily produce false headlines without investigation."

All attempts at counterargument were ignored. The press did not want to hear that Doug and Dave were not and could not have been responsible for all those thousands of far-flung crop circles appearing all over the world for at least two decades. To put the matter into perspective: to accept that Doug and Dave produced all those crop circles is equivalent to finding a couple of counterfeit $20 bills and concluding from this that all $20 bills are counterfeit and have always been counterfeit. Yet, Doug and Dave set the stage for the next ten deeply polarized years of crop circle study up to and including the present.

Maybe this book will help reverse the process. I hope so, and for two good reasons. First, anything that embarrasses and infuriates the priesthood of the Church of Progress has, by definition, a positive value — it's a gauntlet thrown at Seth's feet. Second, I am one of those who feels that some conscious and benign force is attempting to communicate something of consequence to us through these amazing formations — if only we could learn its language. And Linda Howe's book is certainly a valuable step in systematically approaching that problem.

There is some reason for optimism. Despite the willful and deliberate media dismissal, the crop circles continue to show up regularly and on schedule every summer. So, whoever or whatever may be responsible for them is not discouraged. Maybe one lesson is that we shouldn't be discouraged either by the perfectly predictable opposition of the media and the scientific community, or by our present inability to understand the origins and significance of the crop circles. Sooner or later, if we persist, Seth will be brought to his knees. Sooner or later, if the crop circles really are talking a sacred language to us, we will learn it. And it might be that we have to learn the language first if we are to disempower Seth — especially in his ubiquitous jackass and scorpion forms.

John Anthony West
Author, *Serpent in the Sky*
Athens, New York
June 21, 2000

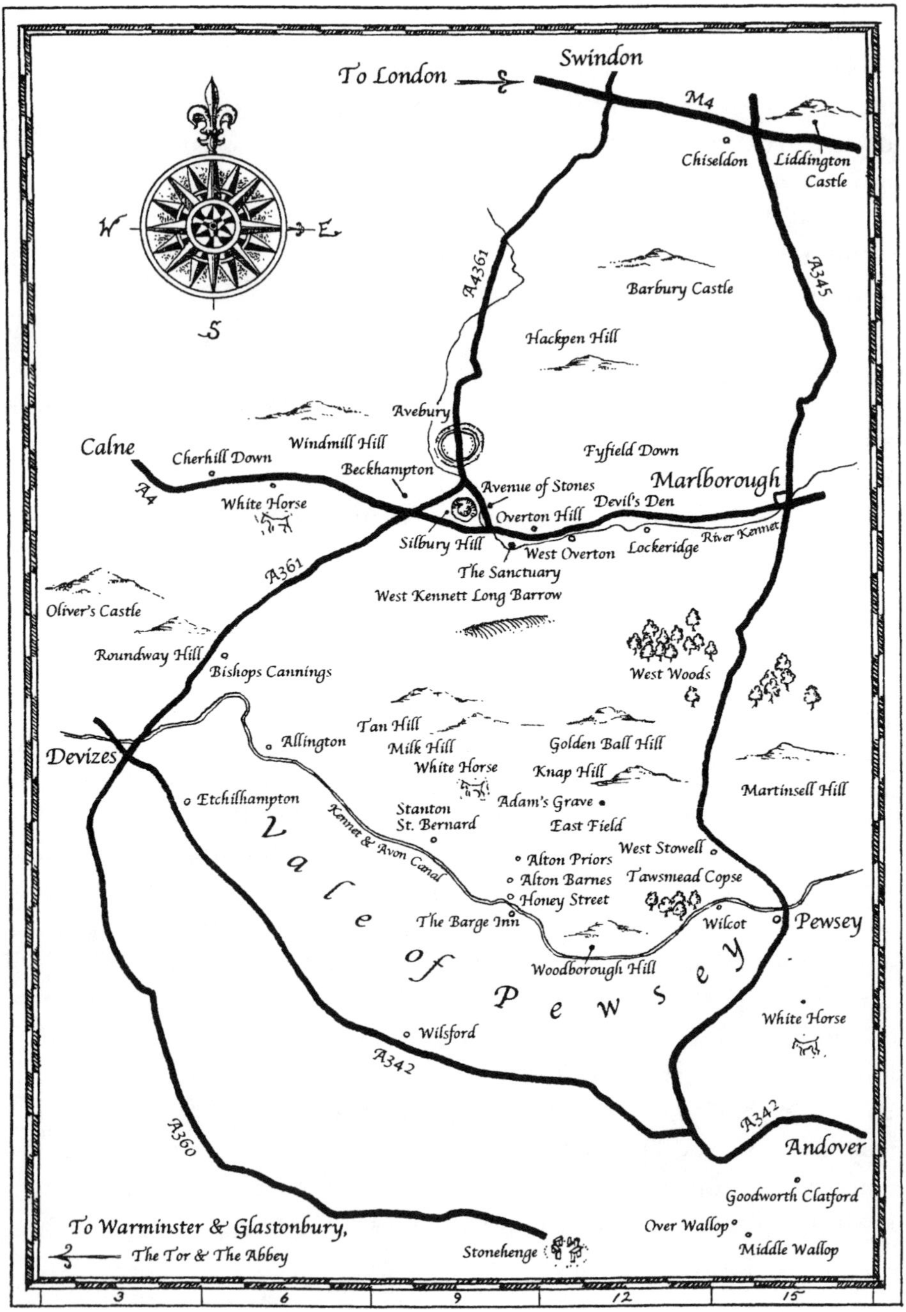

3 miles to 1 inch (approx.)

Map not to scale

I

The Haunted Fields of Wiltshire

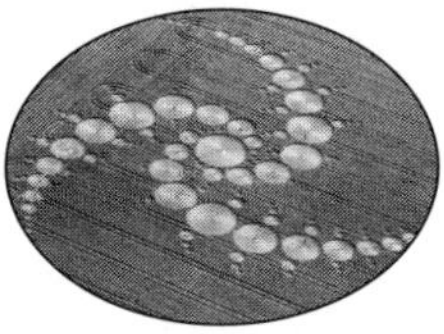

The electronic message said, "A greeting card is waiting for you in one of your favorite spots!" The sender was Peter Sorensen, longtime friend from California and videographer of English crop circles. It was July 20, 1999. Peter had attached one of his digital video frames of a beautiful crop formation discovered on July 18 at Cherhill Down west of Avebury and Silbury Hill. He knew I was headed back to England on July 29 and that a previous 1993 Cherhill formation had affected me deeply.

Peter was working his eighth summer in Wiltshire County about two hours west of London photographing patterns that kept evolving in design and complexity each growing season. Mysterious circles in the British Isles are remembered by adults today who played in them as children in the 1930s and 1940s. No one knew what caused them. The modern age of more complicated patterns and media coverage began in 1981 when a triplet of circles was discovered at Cheesefoot Head, Hampshire.

By the late 1980s, simple circles, concentric rings, quincunxes and "Celtic crosses" of four circles linked by one large, thin ring were photographed and discussed in the international media. The September 18, 1989, issue of *Time* magazine showed large formations in southern England and asked, "Who — or what — is creating those mysterious crop-field rings?"

In the summer of 1991, the patterns were huge and all over the world. Even watery rice paddies in Japan had been imprinted with odd symbols. The term "crop circles" wasn't comprehensive enough and other labels emerged, such as "pictograms" and "agriglyphs." That summer, I was creator and supervising producer of an hour special for Paramount Studios and the Fox Network in Los Angeles about unusual phenomena, including worldwide crop formations. I was in contact with researchers studying soils, plant cell changes, electromagnetic anomalies, strange sounds and lights — all associated with the sudden appearance of unexplained patterns in cereal crops.

Cherhill Down, discovered July 18, 1999, in wheat, 69 meters (226.38 feet) in diameter. Aerial videograph © 1999 by Peter R. Sorensen.

The simultaneous international scope of the phenomenon baffled everyone and contradicted one TV tabloid effort in September 1991 to explain it all away as the work of two elderly gentlemen in southern England. Those studying the facts knew the TV story was not true and were frustrated that the global media would so easily produce false headlines without investigation.

Earlier, in spring 1999, Peter had e-mailed some of his other photographs of April and May formations that could be interpreted as a sun or moon eclipse motif. The sun blackened at noon was on my mind because friends had asked me to join them in Reims, France, for the August 11, 1999, total solar eclipse, the last of the 20th century in Europe.

If the weather was good, it would be the fulfillment of a wish I'd had since childhood to see the sun covered completely by the moon. The fact that the 1999 spring formations started in England with what could be interpreted as an eclipse motif provoked many to wonder if something extraordinary might happen on August 11.

On top of that, a 19-year-old Dutch teenager named Robbert in Hoeven, Holland, had reported seeing a light "make" a 30-foot circle around 12:15 AM on June 7, 1999. The family wanted anonymity, but also wanted to understand what was happening and welcomed research by Nancy Talbott, organizer of the BLT Research Team in Cambridge, Massachusetts. "B" stands for John Burke, researcher; "L" for W. C. Levengood, a biophysicist. "T" is Talbott.

Dr. Levengood has published over fifty scientific papers in peer-reviewed journals. He holds a dozen patents, including a seed analyzer. His crop formation research began in 1991 after he received plants and soil at his private Michigan laboratory. Nancy has focused on organizing plant and soil sampling from formations and on finding grant support for Dr. Levengood's research. After a decade of studying plants and soil from at least three hundred formations in England, Canada, Israel, the United States and other countries, he has concluded, "Whatever is doing these formations is affecting the fundamental biochemistry and biophysics of the plants."

Since 1997, Robbert has seen unusual lights in the fields around his house associated with circles or ovals. Nancy had visited his home several times to collect plants and soil. But this June 7, 1999, event was more dramatic and Nancy called me the next day. We recorded an interview for the news segments I report on the nationally syndicated Premiere Radio Networks.

In that interview, Nancy told me Robbert "opened the curtains and looked out. His window overlooks this little garden out back and the huge farm field which is directly behind their garden. And what he saw is just incredible! He described a small, misty, pinkish-purple light. It was not a sphere, more football-shaped, but slightly smaller. The light was coming in over the field at a height of about ten feet. He watched it as it came closer and closer to a spot about 150 feet behind his house and just stopped. The word he used was 'hanging' in the air. Then the light started to elongate, spreading out, 'becoming thinner and thinner looking like a disc' is how he described it.

"When it was about thirty feet in diameter still hanging there,

Robbert saw what he calls 'electrical discharges' from the bottom of the pinkish-purple disc. Then the light phenomenon simply dissipated and was gone. Robbert was very excited and ran out into the field. Lo and behold, right smack where that thing had been was a 30-foot-diameter circle in wheat. And a 10-foot circle very nearby. He knew the circles weren't there when he went to bed at 10 PM because in the summer it's still light in Holland."

A few weeks later, Robbert saw more lights in the field and ran to find another crop formation that was crackling with electrostatic sounds. When he reached down to touch the soil and plants, they were *hot.*

Field behind Robbert's house in Hoeven, Holland, late December 1999, had a brown discoloration where a mysterious light and circle had been in June. Photograph © 1999 by Bert Janssen.

In 1997, Robbert had been riding his bike through a crop field when he saw hundreds of glowing spheres the size of tennis balls and grapefruit. Some approached and began circling around him. The next thing he knew, he was waking up on the ground in the middle of a brand new circle.

Nancy described his impression of the circle makers. "Robbert said the crop formations are very important, somehow connected to what he thinks are angels, and humans should be paying attention to them. It has something to do with problems in the environment and *deceit.* The crop circles, he says, are like an 'antidote,' but he has no idea what exactly that means or why he is involved."

Once Robbert described a female figure he thought was either associated with one of the lights or transformed from one of the lights. In my research, I found one unusual account of a ball of light that took on human form. The report is in a 1989 book, *Unexplained Mysteries of the 20th Century,* by Janet and Colin Bord. They write, "The witness was climbing in the Bavarian Alps of West Germany around 1950 and she got lost. 'All of a sudden, I noticed a sort of a big ball of light and this condensed into the shape of a tall, rather Chinese-looking gentleman....He bowed, spoke a few words, led me by a small path to the tourists' way and disappeared back as a ball of light.'"

Janet and Colin Bord asked, "Is there perhaps a life form unknown to us which can make itself visible as a ball of light?"

Other increasing evidence suggests that low-energy plasmas are involved. Are their energy and organization a natural phenomenon not yet acknowledged or understood by mainstream science? Could plasmas and microwave frequencies be purposefully manipulated by human and nonhuman intelligences to interact with crops?

What *are* the mysterious lights associated with crop formations since at least 1990, including ones invisible to human eyes but visible on film and videotape? Why does it seem that more people are seeing odd lights above the fields, even photographing them? What was it about the 1990s that provoked the lights and formations to appear around the world, especially in southern England? If U. S. and British military were experimenting with satellite or aerial drone targeting for advanced weapon systems, as some suggested, why would officials subject their own citizenry, crops and animals to potentially dangerous energy beams?

The last year of the 1990s was full of anxiety and anticipation. There were uncertain countdowns. One was the August 11 total solar eclipse. Totality would not be seen in England again until 2090. Solar and lunar eclipses have always been important in the British Isles, where Neolithic and Celtic peoples constructed thousands of stone circles all over Ireland, Scotland, Wales and England to track cycles in the sun, moon and stars. Would the weather be clear enough on August 11 to see the noon sun go black?

Later, on November 16 to 17, the peak of the Leonid meteor shower might reach several hundred or a thousand pieces of rock and dust per hour flashing through the sky at the peak of its thirty-three-year cycle. Would the hundreds of communications and imaging satellites orbiting the earth survive an intense meteor shower without some damage? And Y2K — would the world's computers shut down at 2000 and bring chaos?

Whatever else was going to happen, I wanted a chance to see that solar eclipse and to be in England again, walking along tractor paths known as tramlines with crops all around me waist-high, moving toward a new formation and the mysterious and palpable atmosphere that clings to some for days. I also wanted to interview eyewitnesses of the mysterious lights associated with crop formations.

On earlier trips to England in 1992, 1993 and 1995, I had walked in many formations. But one that stood out was on August 8, 1993, at Cherhill Down. There were no signs of entry on that early morning of its discovery. Four of us were the first to survey, and we had all become light-headed. Headaches followed. We wondered if a residual energy were still affecting the soil, plants and air around us. Since the phenomenon of large pictograms had begun in 1990, many people had reported physical effects inside crop formations — some negative, some positive. There is also the issue of possible allergic reactions to pesticide sprays.

Dr. Levengood had reproduced some of the biochemical and biophysical effects he found in formation plants by exposing normal control plants to various microwave oven frequencies for a few seconds. Any longer exposure cooked the plants. Also, he repeatedly found melted iron oxide particles in crop formation soils. Microwaves could melt iron. The natural level of iron in soil is 0.4 milligram per gram. Dr. Levengood has found *10 to 300* milligrams of iron particles per gram of formation soil. Its distribution has been neatly plotted on a graph consistent with the laws of centrifugal force.

Further, he found seeds from some formations that were stunted in development while others were enhanced and produced accelerated growth after planting. So Dr. Levengood focused research on free radicals that could be produced by microwave energies. Free radicals would interfere with metabolism and weaken or kill plants. (See Chapter 8.)

Dr. Levengood wondered if plants try to counterbalance free radicals and trigger the production of more antioxidants. Increased antioxidant production might explain the enhanced vigor, accelerated growth and resistance to light and water deprivation he has found in some crop formations. The counterbalance between free radicals produced by microwave energies and antioxidants produced by affected plants might depend upon which stage in the plant growth cycle was impacted by the outside energy.

After studying these differences for several years, Dr. Levengood hypothesized that a "spinning plasma vortex of unknown origin containing microwave and ion-electron pulses" was primarily responsible for the physical and chemical changes he has documented. He did not find these changes in control studies where circles had been deliberately made with boards, strings and lawn rollers. Dr. Levengood hypothesized that when people got dizzy, developed headaches or even felt a sense of euphoria, the sensations could be related to residual changes in the local, electric and magnetic field brought on by the spinning ions. But what was the source of the spinning plasma vortex? Was there a connection between plasmas and the lights?

The night of August 7–8, 1993, eyewitnesses reported seeing odd aerial lights near Cherhill. Later, in another nearby field, Peter Sorensen discovered soil and wheat that were rust-colored. When Dr. Levengood received samples at his lab to study, he found thin coatings of iron oxides, magnetite and hematite deposited on the wheat heads, stems and soil.

He wondered if the spinning plasma vortex energies might have come down from a high altitude through iron-containing dust associated with meteors entering our planet's atmosphere. Dr. Levengood speculated that as the spinning energies created the formation, captured iron particles were deposited in molten form onto the plants and soil.

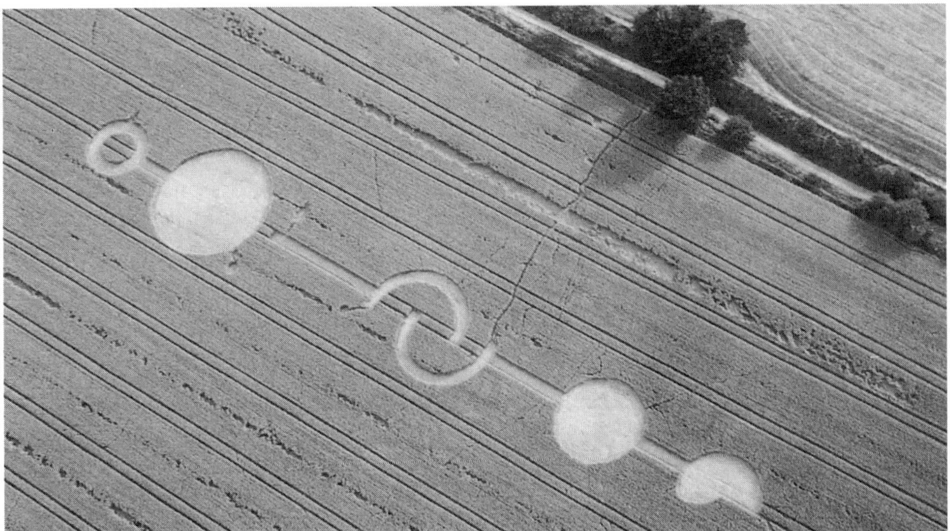

Cherhill Down, 445 feet in wheat, August 8, 1993.
Aerial photograph © 1993 Colin Andrews.

The overlapping arcs in the 1993 Cherhill formation intrigued me. Later, looking for symbol matches, I studied *The Civilization of the Goddess, The World of Old Europe* by Marija Gimbutas. I learned the overlapping arcs are found in an Old European script symbol called "lu" that in some translations mean "light." The sacred writing was used between 5300 and 4300 B.C. to inscribe religious objects and to communicate between man and the gods.

I had the "lu" symbol and Cherhill on my mind when the plane landed at Heathrow Airport in London on Friday morning, July 30, 1999. I was thinking about Robbert's recent light encounters and a quote I had used in my two-volume book, *Glimpses of Other Realities.* Theoretical physicist David Bohm had contemplated the relationship between energy and mass and wrote, "Mass is a phenomenon of connecting light rays which go back and forth, sort of freezing them into a pattern. So matter, as it were, is condensed or frozen light....You could say that when we come to light, we are coming to the fundamental activity in which existence has its ground, or at least coming close to it." That's what Albert Einstein's famous formula $E=mc^2$ was about. Energy equals mass times the speed of light squared. If you put m on the left side of the equation, E moves to the right side over c^2. Then it reads $m=E/c^2$. Mass equals energy and light. Or as Bohm put it, matter — you, me, the crops, the birds and the stars — is condensed or frozen light.

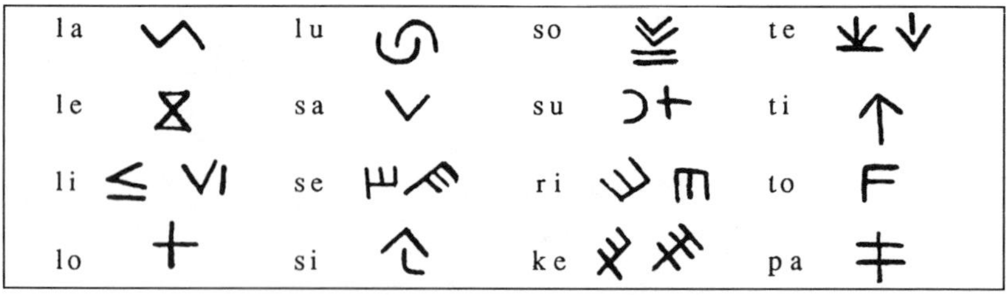

Symbol "lu" in Old European script, 5300–4300 B.C. From The Civilization of the Goddess, The World of Old Europe, *© 1989 Marija Gimbutas.*

Now I was about to reenter the hauntingly beautiful county of Wiltshire, where mysterious flickering lights and glowing spheres are often seen, where ancient grave mounds called tumuli have even contained a few skeletons connected to Mesopotamia, where hundreds of weathered stone circles are linked to sun and moonlight and the modern ones in cereal crops are linked to unidentified lights. Are the crop lights new? Or as old as the ancient sarsen circles? If they are plasmas, what organizes them into energy systems that can create complex formations in cereal crops, grasses, dirt and even snow? People who have encountered glowing spheres sense an intelligence. But what is it?

I awkwardly got into the right side of my British rental car and drove around the parking lot several times to readjust my thinking so I could stay on the left side of the road instead of the American right. Judging the distance to curbs, rocks and walls on the left from the unfamiliar far right driving position is difficult, and British gas stations are used to repairing left flat tires on rentals.

Eventually I had the courage to turn right out of the Avis lot into the left lane and kept following signs to the M-4. That main freeway across southern England has six lanes of cars and trucks moving to and from London. The M-4 cuts through vast fields of wheat and other cereal crops not far outside the city. The beautiful farmland flows up and down hills, unchanging year to year, century to century, without any intrusion of modern billboards or neon signs.

That close juxtaposition of the fast, noisy, moving cars to the huge, quiet expanses of grain is a paradox, like England at the end of the 20th century. While the supersonic Concorde flies overhead getting passengers from London to New York in two hours, a soldier's boots from World War II still hang above a pub table in Orford on the sea between England and the Netherlands. "They'll always be there for him," the pub owner said, cutting to the heart of England's reverence for history and sacred sites.

Ninety minutes west of London, there's an exit off the M-4 to Marlborough. Gearshifting down from the speed of six lanes to two, there is also a shift from the urgency of huge trucks trying to make deadlines to the quiet of a narrow country road. After miles of straight, flat asphalt, a calm takes over in the shadowy tree tunnels and villages of thatched roofs, pubs decorated with royal coats of arms and baskets of red, blue, yellow and pink flowers near pastures of grazing sheep. It's Wiltshire County, where time seems to stop. The stone circles, monoliths and tumuli mark sacred sites that are thousands of years old.

On July 30, 1999, I took that Marlborough exit and headed straight for Cherhill about three miles west of the ancient and sacred stone circle and avenue of Avebury, the largest in the British Isles. Nearby is Windmill Hill, where the astonishing Triple Julia Set of 194 circles spanning a thousand feet in diameter was found on July 29, 1996. Windmill Hill was in use around 3700 B.C. for Neolithic rituals with Silbury Hill and the Sanctuary at Avebury that involved cycles of the sky, fertility, birth, life and death. Silbury Hill is the tallest artificial mound in Europe, 130 feet high, built for purposes unknown. But 20th-century researchers such as Moses B. Cotsworth from Yorkshire, England, studied Egyptian pyramids and postulated that at least one Silbury Hill function was as a clock/calendar with a pole on top to cast shadow markers through the seasons.

Further, Professor Alexander Thom, chair of engineering science at Oxford, wrote a 1967 book entitled *Megalithic Sites in Britain* in which he showed how the stone and wood circles or henges of Britain in the second millennium B.C. were aligned on certain stars. Thom also determined that the ancient stone henges used a geometry as sophisticated as later discoveries by Pythagoras. Thom said the ancient stone circles were uniformly built on a unit of measure that he called a megalithic yard of 2.72 feet, or .829 meter. Thom was convinced megalithic sites in Britain were calendars and clocks based on bright rising and setting stars.

Thom said the ancient engineers built perhaps 10,000 megaliths from one end of Britain to the other with great accuracy in alignment. One Stonehenge expert, R. J. C. Atkinson, professor of archaeology at University College in Cardiff, England, agreed with Thom's data and marvelled at the high degree of competence in empirical astronomy that existed in Britain more than 4,000 years ago.

As I drove toward Cherhill again on July 30, 1999, I visualized the 1993 pictogram with its overlapping central arcs that once meant light and wondered why the same fields were used over and over by the circle makers. Was there something about the specific geography, all those megalithic stone sites, that created a connection to the unseen that no human understood? Was there a specific connection to the Celts and their Druid priests and to the inspiring intelligence behind the Mesopotamian and Egyptian civilizations? William Stukeley, a clergyman and Freemason, studied comparative religions and wrote, "The plan on which Avebury was built is that sacred hierogram of the Egyptians and other ancient nations, the circle and the snake. The whole figure is the circle, snake and wings." Many historians have disagreed with him, but his hypothesis persists, especially since the recent discovery of another stone avenue at Avebury that tends to reinforce Stukeley's concept.

On June 19, 1999, a 670-foot-long formation emerged in a wheat field near Silbury Hill that reminded some of the "Winged Sun Disk" symbolic of the Egyptian sun god, Ra.

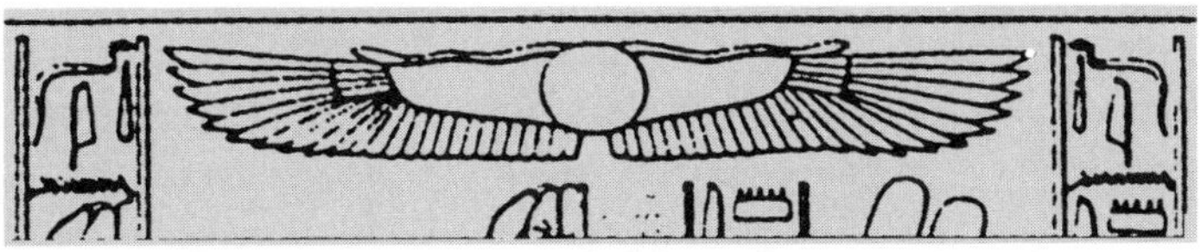

Ra, Egyptian god depicted as Winged Solar Disk, carved in the center of the west wall sanctuary at the Temple of Luxor. From Serpent in the Sky, The High Wisdom of Ancient Egypt, © *1993 by John Anthony West.*

The Egyptians honored Ra as ruler not only of the sun but of all the sky. Ra was revered by the Pharaohs, who considered themselves his sons. Ra was said to reside in the ancient city of Heliopolis where he was worshiped in the form of an obelisk believed to be a petrified ray of the sun.

Egyptologist and mathematician R.A. Schwaller de Lubicz said about the Ra solar disc in his 1957 masterwork, *The Temple of Man,* "Every symbol, even one beyond interpretation, is in itself directly related to the thing that it expresses…or evokes. Thus, wings can be nothing other than what evokes the fact of flying. That which flies can be nothing other than what rises and escapes the earth.…Symbolism both conceals and teaches, and the more complicated the teaching becomes, the more it conceals and removes one from knowledge."

If, for a moment, we assume that humans are not the easy explanation for all that has occurred in the world's mysterious crop formations, and that plasma energies can be created and manipulated by intelligences other than Mother Nature, we might perceive a pattern. We might learn that a formation at Cherhill in 1993 and another in 1999 were somehow connected in symbolic meaning, or hidden code, even if the patterns are different. I have received off-the-record information from more than one source that U. S. and British intelligence have been photographing crop formations from satellites, trying to assemble a chronological database for computers to analyze.

Sometimes as many as ten formations were reported on the same day and in different counties, as happened on Thursday, July 29, 1999, according to the www.cropcircleconnector.com's 1999 Archive. During one eight-day period between June 12 and June 19, 1999, sixteen formations were reported in seven counties of southern England, five on the same night of June 19.

"Winged Solar Disk." 670 feet long, discovered in wheat field near Silbury Hill, Wiltshire, June 19, 1999. Aerial photograph © 1999 by Lucy Pringle.

Two of the longest crop formations in 1999, 1,020 feet and 500 feet, came down in the East Field at Alton Barnes between midnight and 5 AM on June 12. One man camped in the field was shocked to wake up and see the enormous patterns right in front of him.

"I'm convinced absolutely 100 percent that no one else had been in those crop formations during the night," he said, because he was there. (See Chapter 6.) That same night, two other formations also occurred in Chiseldon and All Cannings in Wiltshire.

Sometimes I wonder if crop formations were laid down in moves and countermoves like a chess game played by unseen competing forces. Robbert said deceit was involved and formations were an "antidote." One ancient code system used symbols called sigils to disguise names and locations of people, angels and angelic realms in that old struggle of light and dark. The sigils were odd patterns of circles, loops, crosses and other marks. An example are the sigils for Raphael and Ophiel on page 17.

In the Greek and Hebrew alphabets, each character stands not only for a sound but also for a numerical equivalent. The key to using sigils was to know the number and letter code beneath the deceptive overlay of circles, loops, crosses and other marks. Angel names were also assigned sigils in order to secretly contact them in white and black magic rituals.

Perhaps historic forces continue to prod and provoke humanity in unexpected ways. Huge stone structures seem to echo a forgotten past that nags our subconscious. Maybe that's why I was happy to see again that "petrified ray" rising from Cherhill's chalk hills. It was copied after the classic Egyptian elongated pyramid and placed high as a memorial to all the English lives lost in World War I.

Cherhill Down, July 30, 1999, photo from roadside twelve days after formation appeared. A path trudged through the wheat by tourists led to the nine-pointed star directly below the obelisk. The white chalk horse is to the left. Photograph © 1999 by author.

Not far from the obelisk is one of nine White Horses that have been carved into Wiltshire's Upper Cretaceous Chalk, which is so permeable it resists erosion. Horses were honored by ancients as both spirits of the solar cycle and as vehicles that carried the souls of the dead to the Underworld. But no one is certain why the White Horses were first carved in the chalk hills. The oldest and perhaps most elegant, at Uffington, goes back about 5,000 years. The one at Cherhill was carved in 1780. In the 1990s, many of the finest crop formations have been in fields below White Horses.

Even though I arrived at the 1999 Cherhill formation two weeks after it first appeared, the fluid lay of the wheat in several overlapping directions was still clear. There was also a small symbol a few feet beyond the western side of the formation that some referred to as a "signature" or seal. In early Mesopotamia, seals of geometric and other designs were the earliest symbolic representation of individuals or groups that linked directly to property or responsibility. Studying the long history of seals, occult sigils and Mystery School symbols is a complex ground upon which all serious crop formation researchers eventually meet.

Cherhill Down, July 30, 1999, wheat flowing in two directions around one of nine triangles surrounding six crescents below the chalk White Horse carved in 1780. Photograph © 1999 by author.

Cherhill Down formation, 69 meters (226.38 feet) in diameter, discovered July 18, 1999, showing small symbol "signature" on right. Aerial photograph © 1999 by Steve Alexander.

That 1999 Cherhill signature, or seal, matched another from the year before at the ten-pointed star composed of one five-pointed star laid on top of another, discovered in Dadford, Buckinghamshire, on July 4, 1998. Three small symbols were detached from the main formation. In the aerial photograph on the next page, the symbol at far right is an Egyptian ankh, or ansate cross, representing life and reincarnation.

In the Mystery Schools of Egypt, the student passed through all forms of actual and imaginary dangers holding the Crux Ansata above his or her head to make the powers of darkness fall back. The middle signature matched the one in Cherhill 1999 and was similar, although not identical, to the signature at the end of the long East Field "Sentence" of 1999 in Chapter 6. The third symbol on the left was an upside-down combination of a Tau (T-bar) and a circle.

Dadford, Buckinghamshire, formation found in wheat, July 4, 1998, 175-foot diameter, with three "signatures." Center signature matches Cherhill 1999 and is similar, but not identical to, the East Field "Sentence" discussed in Chapter 6. The Tau T-bar and circle on left is similar to a pentacle symbol for planetary angels reprinted by Manly Palmer Hall in 1928. Aerial photograph © 1998 by Steve Alexander.

I began researching symbols and found one provocative similarity in "Pentacles of the Seven Planets and the Seals and Characters of the Planetary Angels" from an unpublished medieval Book of Spirits printed in *The Secret Teachings of All Ages: An Encyclopedic Outline of Masonic, Hermetic, Qabbalistic and Rosicrucian Symbolical Philosophy* (1928) by Manly Palmer Hall. A Tau (T-bar) combined with a Taurus-like symbol is shown above an encircled triangle identified as a pentacle for "planetary angels" Raphael, Uiriel, Ophiel and Cochabiel.

As discussed earlier, these symbol "addresses" can be used to call upon powers in the unseen. The triangle within a ring and circle described as a pentacle in the Hall book is not typical. Traditionally, a pentacle or pentagram is a five-pointed star formed by five straight lines connecting the vertices of a pentagon and enclosing another pentagon in the completed figure.

In the unpublished reprint in Hall's book, the Tau/Taurus represents the pentacle for one of seven planets controlled by the angels Raphael, Uiriel, Cochabiel, and Ophiel. The sigils for Raphael and Ophiel are contained within the two smaller circles.

From a mediæval *Book of Spirits* (unpublished).

THE PENTACLES OF THE SEVEN PLANETS AND THE SEALS AND CHARACTERS OF THE PLANETARY ANGELS.

The seven large circles are the pentacles of the planets, while the two small circles under each contain the seal and the character of the controlling intelligence of the planet.

One of seven pentacles that represents a planet controlled by the angels Raphael, Uiriel, Cochabiel and Ophiel. The sigils for Raphael and Ophiel are contained within the two smaller circles. From The Secret Teachings of All Ages, *page 104, © 1928 Manly Palmer Hall, published by the Philosophical Research Society, Los Angeles, California.*

Pentagram, or pentacle, which is endless like a circle. At its center is a five-sided pentagon. Drawing © 2000 by author.

According to *The Secret Language of Symbols* by David Fontana, the pentagram, or pentacle, "being endless like a circle,... stands for perfection and wholeness. At the same time, its (five) triangular components stand for the four elements, plus spirituality. These twin properties give the symbol power to bind or banish evil spirits and make it a favorite of magicians."

A variation on the endless concept is the Seal of Solomon, which goes back thousands of years. Fontana says it "appears in the writings and practices of magicians, and has strong associations with Hebrew mysticism: hence its appearance as the Star of David." Jewish scholars have discovered that the origin of the woven hexagram symbol is probably Arabic.

The upward-pointing triangle is equated with fire and masculine energy; the downward-pointing triangle to water and feminine energy. Fontana says, "The base of each triangle bisects the other just under the apex," forming two more symbols for air and earth. "The Seal therefore portrays the four elements, which are combined with the number three (the triangle) to produce the magical number seven. However, as the Seal only has six points, the seventh is said to be invisible, representing the spiritual element of transformation, which emerges to the inner eye of the magician as he meditates upon the Seal."

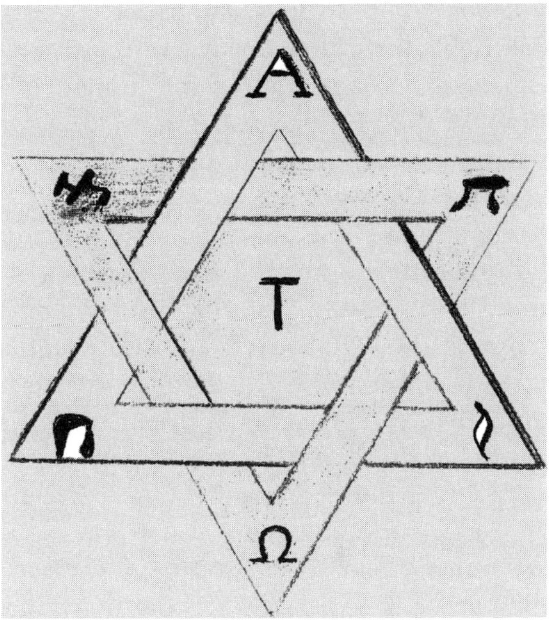

The very ancient Seal of Solomon, a six-pointed star formed by extending each of the sides of a regular hexagon into equilateral woven triangles, likely of Arabic origin. Alpha and Omega are top and bottom points; Tau is centered within the six-sided interior hexagon. Drawing © 2000 by author.

A variation on the pentagram and the July 4, 1998, Dadford formation in wheat was discovered on August 8, 1998, at Beckhampton, Hampshire, about 200 feet in diameter.

Beckhampton, Hampshire, crop formation in wheat, about 200 feet in diameter, August 8, 1998. Diagram © 1998 by Peter R. Sorensen.

The Rosicrucians combine Egyptian Hermetic writings, Christian Gnosticism, Jewish Kabbalism, alchemy and other Mystery School beliefs and practices. Hermetic works are about the human relationship to the cosmos as described by the Egyptian god Thoth, known to the Greeks as Hermes. Thoth/Hermes was believed to be the original source of writing and divine wisdom.

The fundamental concepts in ancient Hermetic writings and Rosicrucianism are that the cosmos is a unity of which all parts are interdependent within it. Further, if humans can understand and use the natural laws of the cosmos, they will learn to transmute common metals into gold and mortality into immortality. It was necessary to know the laws of sympathy and antipathy by which parts of the universe were related. The yin and yang symbol represents that tension of opposites within the whole. Another Mystery School concept was the recycling of souls as the machinery of the universe. Souls emanated from the Source, fell into bodies and had to return home to the Source.

Over the 1989 to 1999 decade, there were more than two thousand formations worldwide, but only a few "signature" symbols. Another signature in 1999 was found at Liddington Castle, nine miles north of Avebury, on July 21. The concept of a signature implies an intelligence that claims or marks its creations. As this story unfolds, it will be clear that hoaxers are not the explanation for all the complex formations and signatures.

Liddington Castle formation in wheat, 63 meters (207 feet) in diameter, discovered July 21, 1999, with a "signature" at upper left. Aerial photograph © 1999 by Peter R. Sorensen.

As already discussed, and with more detail in Chapter 8, scientific studies have confirmed metabolic changes in some formation plants that cannot be hoaxed. This research contradicts Doug Bowers and Dave Chorley from southern England, who told a London tabloid TV reporter in September 1991 that they were the answer to the crop formation mystery. Their proof, they said, were their initials, "DD," laid down next to circles, insectograms and other patterns in two southern British counties in the early 1990s. Inexplicably, the international media accepted their sudden claim and headlined that the crop mystery had been solved. But the letters "DD" are also carved beneath a T-bar on an ancient Celtic stone in Lancashire, England. John Michell, first editor of *The Cereologist*, printed a photograph of the stone in 1992. They weren't Doug and Dave's initials, but what do they mean?

Ancient Celtic stone in Lancashire, England, with what appears to be "DD" beneath a T-bar amid other symbols. Photograph © 1992 by Dominic Morton and printed in The Cereologist, *Vol. 7, 1992.*

Why didn't the news media investigate the legitimacy of the two men's claims? If there had been any substantive research, reporters would have learned that in the summer of 1991, complex crop formations were reported around the world, including rice paddies in Japan and fields in Australia, Canada, the United States, Russia and many parts of Europe beyond southern England.

By 1993, Doug and Dave said they had stopped rolling down crops, and Dave Chorley died in the fall of 1996. But the formations continued with increasing complexity in design and physical impacts on plants and surrounding soil that could be studied. Doug and Dave were not the answer to the worldwide phenomenon and never had been.

Debates have raged about who or what the circle makers might really be. Human hoaxers have performed before television cameras using equipment and spending more than twenty man-hours to reproduce rough copies of some formation patterns. However, no human or group of humans has produced a 1,000-foot-diameter formation of hundreds of circles in the short summer hours of darkness.

Speculations about this global phenomenon have included meteorological events, military satellite experiments, organized plasmas, earth subterranean Devic energies, extraterrestrial calling cards and other dimensional communications, including time travelers.

The ET speculation was fueled in the early 1990s by eyewitness observations of unidentified lights near fields before or after a formation's creation. Skeptics offered military flares and lighted blimps as explanations, but neither of these technologies can move away rapidly, nor can they disappear in one spot and reappear seconds later in another part of the sky. The lights were most often described as not having supporting or surrounding structure, even when seen through binoculars.

On the night of July 16–17, 1991, when the extraordinary Barbury Castle triangle was formed, eyewitnesses reported seeing many odd lights; a huge, black aerial triangle that blocked out stars; and even a *curved* beam. (See Chapter 2.)

By the summer of 1999, researchers and European residents who have followed the mystery for several years were speculating more about other dimensions, parallel universes and time travelers. But why would crop fields be the medium? And why so widespread in England, France, the Czech Republic, Germany, the Netherlands, Spain, Italy, Japan, Australia, the United States, Canada, Russia, Peru and other countries?

Who, how, and why are questions that provoke any mind that has seriously studied the formations, their repetitive geometries and fractal themes, the biochemical and biophysical changes not found in known hoaxes that include heating by microwave energies; increased vigor and accelerated growth in some plants and less vigor and retarded growth in others; magnetic anomalies in soils; and disturbances in cameras, tape recorders and batteries that cause equipment to stop operating.

What do the mysterious lights and spheres look like? Are there any unusual sounds? What physical and chemical changes occur in affected plants and soil? These questions are answered by many eyewitnesses, their photographs, videotapes and drawings, plus scientific analysis.

But there are other questions not yet answered. If hypothetical spinning plasma vortices are involved with crop circle creations, what unknown trigger would repeatedly produce them? Are sound frequencies applied, perhaps outside the range of human hearing? Could low-energy plasmas with microwave frequencies produce the physical impacts on soil and plants and leave images of mysterious lights and spheres on videotapes and photographs? Are the mysterious lights an unidentified natural phenomenon? Or are they supervised by an advanced intelligence that humans have yet to encounter? Could the lights be connected to the same ancient intelligence that inspired other marks around the world in ziggurats, pyramids, mounds and stone circles?

2

Strange Sounds and Flickering Brightness

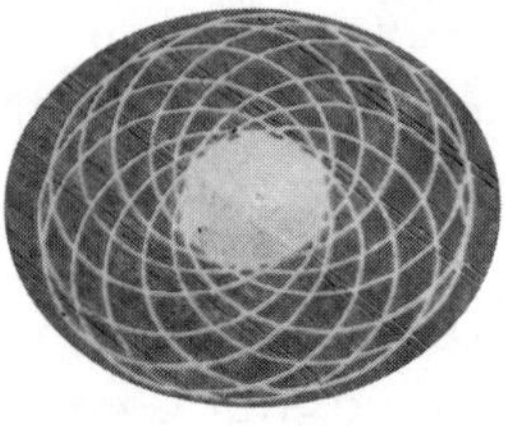

In June 1989, English crop circle researcher Colin Andrews had flown in a helicopter with a CBC Canadian television crew to film eleven crop circles adjacent to Silbury Hill. Later that night a local resident saw a very unusual light descend from the sky to those same circles. According to the eyewitness, after midnight he walked into his backyard and saw a bright orange light he estimated to be about 30 feet in diameter moving vertically down toward the field in front of Silbury Hill. The man could see a tree on the boundary line of the wheat field silhouetted between him and the orange light, so he knew the object touched down in the field of eleven circles that the TV crew had filmed only hours before. After the sun rose, the witness could see that a twelfth circle had been added in the field.

The next year, in June 1990, researcher George Wingfield and two other witnesses visiting Wiltshire watched "mysterious lights, very small, which slowly and thoughtfully glided closely over the wheat, perhaps 200 to 300 yards away from us." Wingfield said he watched the small lights move over the crops for an hour, periodically joining together as if in communication and then flying away. When Wingfield tried to approach them, the lights slowly moved away from him.

A month later, on July 26, 1990, photographer Stephen Alexander was atop Milk Hill in the late afternoon with his video camera. While taping a two-week-old pictogram at nearby Stanton St. Bernard, Alexander and his then-wife, Janice, saw a flash in the wheat below.

In his own words, Steve later wrote: "The tape shows a small white light source moving among the heads of the corn (wheat) at an estimated speed of around four to seven miles per hour. Its characteristics are erratic. At some moments its pace is constant. But at others, it appears to flicker or reflect sunlight before slowing down and blinking out of sight. It then reappears, remains motionless, before repeating the same process in a random fashion as it passes through the crop, possibly even cutting a swathe on occasions. In a matter of some thirty seconds, it completes an entire circle, having crossed tractor tramlines at right angles and followed their course on other occasions, all without any possible indication as to the light's origin.

"Pulsing, flashing, flickering and shining are all words to describe this anomalous light which appears to bear a diameter of no more than eight to ten inches. To say the phenomenon seen is spherical would be inappropriate as it seemed to be more a light burst than a solid object.

"At a distance of some 850 yards away, the flickering, pulsating light is seen to leave the wheat field, cross over a trackway and then enter a second field — this one in the process of being plowed out by a tractor."

Months later, Colin Andrews located the farmer who had been driving the tractor. His name is Leon Besant. He described "a silver glinting light" about the "size of a beach ball" that was "flying around in the corn, on top of the corn." Besant compared the light to tin foil blazing in the sun, "a constant glint. I couldn't make it out. I don't know what it was."

Japanese video experts who studied the tape said, "The glinting object appeared to have no material substance other than as a pure source of light. In other words, on blinking out between flashes and bursts, the object ceased to exist."

Steve Alexander's videotape sequence of the mysterious small light lasted about five minutes. Toward the end, it is difficult to see the moving object against the setting sunlight. For most of the time, it appeared to be no more than three to four feet off the ground at the level of the wheat heads. Steve reported that the light gained speed, moving into a second field, and "quickly crossed a third and final field before it began to climb the lower slopes of Clifford's Hill, some 1250 yards [from Steve]. It then ascended the fairly steep hill without any effort whatsoever, although by now its pulsations are becoming less obvious and less frequent. The phenomenon is then seen climbing the earthen henge that marks the boundary of Rybury Camp, the Neolithic encampment crowning the summit of Clifford's Hill, exactly one mile from Milk Hill."

Janice Alexander was heard on the still-running videotape to say, "It's gone. No, it's not. It's going up in the air!" The camera lifts upward, but Steve can be heard replying, "Can't see it on here," meaning through his camera lens. Janice said she could see the "light source climbing into the sky, emitting intermittent flashes before it was finally lost from view."

Later, investigators calculated that the light had covered a distance of about 1,100 yards in 106 seconds, or about 40 feet per second.

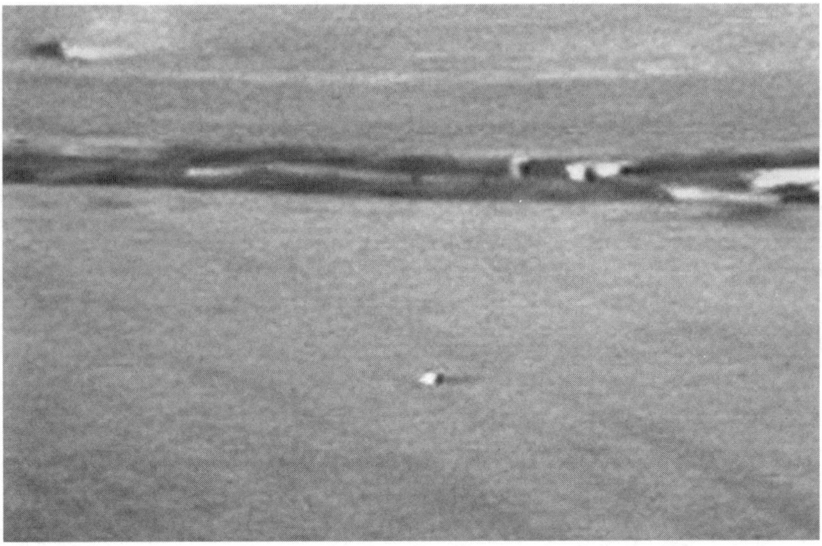

From five-minute video sequence taped July 26, 1990, at 4:30 PM of flickering brightness moving above wheat in Stanton St. Bernard formation below Milk Hill and Clifford's Hill. Video frame © 1990 by Steve Alexander.

Stanton St. Bernard formation near which Steve Alexander videotaped the flickering brightness on July 26, 1990. Photograph © 1990 Timothy Good.

A year later, on August 19, 1991, between 8 and 8:30 PM near Barbury Castle north of Avebury, German students Constantin and Dominic von Durckheim videotaped a similar white, glowing object moving above the extraordinary large triangle that by then was a month old. For several minutes, the white object moved along the top of the crop, dipping periodically down into the plants and then back up before it took off into the air and disappeared.

Video frame © 1991 by Constantin and Dominic von Durckheim, from two-minute video sequence of moving light near Barbury Castle formation taped August 18, 1991, between 8 and 8:30 PM.

Crackling sounds have also been associated with several different crop formations. After Dutch teenager Robbert watched wheat go down beneath a light, he ran from his house about 300 feet into the field. He said the soil and plants were hot to the touch and a strong electrostatic sound came from the freshly flattened plants.

Similarly, in Brentwood, Tennessee, on June 25, 1999, a man and his two sons were surveying a pictogram found in wheat south of Nashville on June 18, 1999. During the hour of their survey work in the 185-foot formation, the three heard a sound they compared to the "snap, crackle and pop of Rice Krispies," or "like static."

"We were surrounded by the sound," Tom Murdic told me in a radio interview. "It was everywhere. Of course, you'd think it was coming from the wheat because that's the only thing in there to make noise. It was like static discharge."

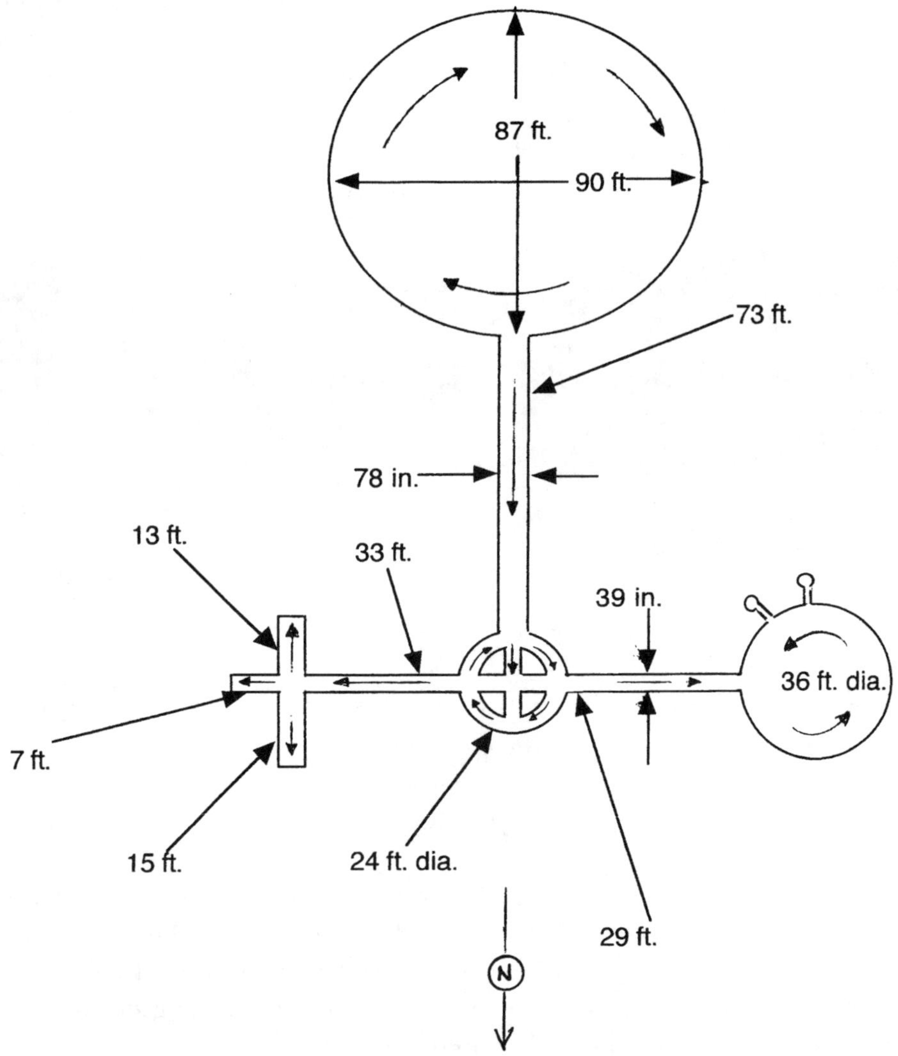

June 20, 1999, survey by Tom Murdic and his two sons. Distance of 185 feet from top of largest circle to bottom of quadrant circle in wheat. Brentwood, Tennessee, south of Nashville. The wheat gave off a sound "like static." Diagram © 1999 by Tom Murdic.

By February 2000, new grass was coming up in the Brentwood, Tennessee, field in the same place and pattern as the June 1999 formation, but in a darker shade of green. Such "ghosts" of the previous year's patterns have occasionally emerged, sometimes in plants that seem to have greater vigor than surrounding nonformation crop.

Brentwood, Tennessee, pictogram in wheat found June 18, 1999. Triangle of small circles in the lower right corner referred to as "grapeshot" by English researchers. Aerial photograph © June 25, 1999 by Tom Murdic.

New grass coming up in Brentwood, Tennessee, field in the same pattern as the 1999 wheat formation, but darker green, February 23, 2000. Photograph © 2000 by Eric Liden.

July 31, 1999 — Wiltshire

That crackling electrostatic sound came again at 6 AM on Saturday, July 31, 1999, at a field in Wiltshire called Roundway — not far from the bed-and-breakfast where I slept my first night back near Cherhill. But I wasn't to know about Roundway for several more hours. At the Barge Inn the night before, the only new formation I learned about had been discovered near Avebury after sunup on July 29. So I got up early to travel there.

On previous visits to England, I had learned how cool and humid summers can be and expected to dress in jeans and a jacket. But this time, even before 9 AM, it was unusually warm, too warm for jeans. Summer 1999 turned out to be one of England's hottest on record, along with many other parts of the planet, as global warming seemed to be turning into a chronic condition, whether natural, manmade or both.

The sky was blue and nearly cloudless as I pulled into the parking lot across the street from the Henge Bookshop in Avebury. I walked the half mile south past the gray sarsen rocks that line up like statues to form the great circle and avenue, which are more than 5,000 years old.

Avebury, England, a Neolithic ceremonial site more than 5,000 years old surrounded by the largest circle of sarsen stones in the British Isles. Formation on hill behind. Photograph © 1999 by Steve Alexander.

Already waiting under an umbrella was a young neighbor hired by the farm's owner, Tony Farthing. Mr. Farthing cultivates 1,100 acres in the fields around Avebury. He had given Robert Hughes a deck chair, an umbrella for shade and a cash box to collect a pound from all who wanted to walk in the formation.

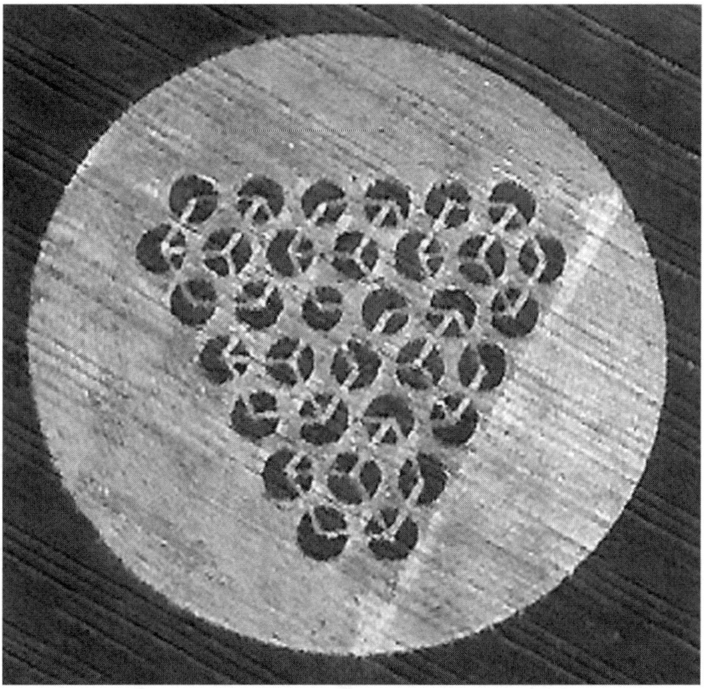

Illusion of six cubes projecting from wheat near Avebury, Wiltshire, England. Circle diameter about 300 feet. Discovered on Thursday, July 29, 1999. Aerial videograph © 1999 by Peter R. Sorensen.

I began by going to the top of the large circle that surrounded confusing patterns of standing crop. Then I walked off approximately half a yard at a step to see how big it was and estimated it at 300 feet in diameter. I walked around the enormous flattened circle and into pathways that ran through the center. The lay of the crop was messy and complicated, not smooth like most others I had seen before.

Robert Hughes showed me an aerial photograph taken the day before. I was surprised to see six "three-dimensional cubes" that seemed to rise upward from the flattened circle.

By August 7, the day I left for France and the solar eclipse, *The Daily Mail* ran a feature article proclaiming the formation was the *Mail*'s commissioned hoax perpetrated by seven men between 11 PM and sunrise with Mr. Farthing's permission. That would be about forty-two man-hours to produce the formation. Canadian researchers Chad and Gwen Deetken later told me they questioned the story because they were in the Avebury Avenue of Stones until 1 AM on July 28, not far from the wheat under a full moon. They thought seven people working on such a large formation would have been seen or heard.

Another visitor to Avebury told Hughes that she had been restless and had walked the Avebury avenue at 3 AM, July 29. She thought all was quiet. The manager of the Henge Bookshop in Avebury told crop researcher Bert Janssen that he was up all night on July 28–29 with windows open facing the hillside field about half a mile away. He neither heard nor saw activity.

Avebury village downhill from the July 29, 1999, six-cube formation in Tony Farthing's wheat field. Photograph © 1999 by author.

Overlapping layers of wheat in Avebury formation, July 30, 1999.
Photograph © 1999 by author.

Tony Farthing told several people and a local newspaper in September that he had been paid by *The Daily Mail* to allow his field to be used by a *Mail* reporter and hoaxers. Mr. Farthing's reason: he assumes the formation creators are human and wants to stop them. Curiously, even hoaxers have admitted over the years that after they have put crop down with boards, rollers and string, sometimes more is mysteriously added to their original design, which spooks them.

Not knowing anything on July 31 about Tony Farthing's alleged collusion with *The Daily Mail*, I was puzzled by the pattern in the aerial photograph when compared with the jumbled lay. But I gathered some samples of the wheat for biophysicist W. C. Levengood to test in his

Michigan lab. Nancy Talbott and another crop researcher, Geoffrey Stray, had also found some unusual wheat in the Avebury formation, in which the stems just below the wheat heads were rippled like Christmas candy, and in others the leaves and stem just below the wheat head were twisted in a unique spiral.

The same crimping and spiraling stem phenomenon was found in wheat sampled from a Whitefish, Montana, formation discovered on August 6, 1999. Dr. Levengood received samples from both Avebury and Whitefish and found biochemical and biophysical changes that he defines as "somatic developmental abnormalities," which are discussed in Chapter 8 of this book. Perhaps his most surprising conclusion is that the energy that affected the plants and caused the developmental abnormalities had to have interacted with the Montana and Avebury fields *at least two to three weeks before* the formations occurred. Are fields premarked and blueprinted in some way? The issue of accelerated and retarded growth is discussed further in Chapter 8.

Scientific Evidence That Cannot Be Hoaxed

Dr. Levengood has analyzed thousands of plants from crop formations because of his curiosity about what unusual energies are interacting with the plants to produce some of the biochemical and biophysical changes he has confirmed in his laboratory. He says that "a spinning plasma vortex with a microwave component and complex lower-energy ion-electron pulses" have interacted with many formation plants he has studied.

He also asserts that the energy interaction that creates formations must occur rapidly, in a few seconds. Otherwise, he would find evidence of burning or charring from the microwave heating. But he does not find such damage, and the formation plants continue to live and grow until harvest.

The biochemical and biophysical changes that Dr. Levengood has documented in thousands of plants from crop formations include a combination of swollen and reoriented growth nodes, ruptured growth nodes, metabolic changes in plant cells and magnetic iron particles deposited over the plant surfaces and surrounding soil.

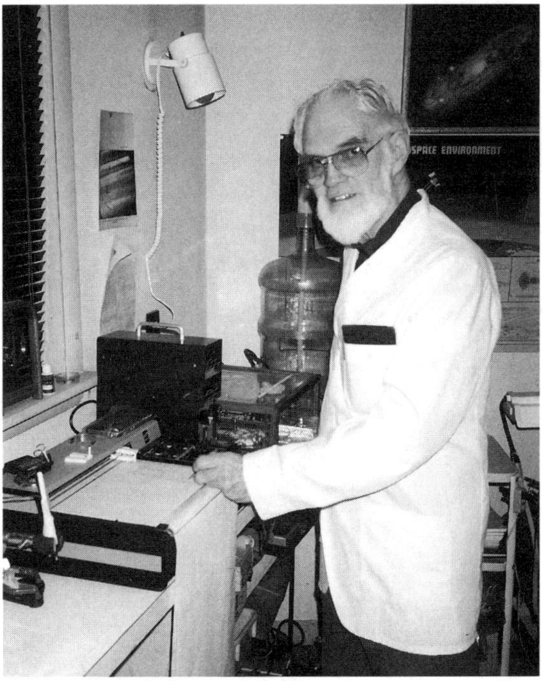

W. C. Levengood, Ph.D., biophysicist,
in his Michigan laboratory. Photograph by author.

He and Nancy Talbott have published his recent research in a 1999 international journal, *Physiologia Plantarum (105:615–624).* They wrote, "The findings reported here lend further support to a decade of research with empirical data which suggests that over 95% of worldwide crop formations involve organized ion plasma vortices that deliver lower atmosphere energy components of sufficient magnitude to produce significant bending, expansion and formation of unique expulsion cavities in plant stem pulvini (swelling at base of leaf stalk), as well as significant changes in seedling development."

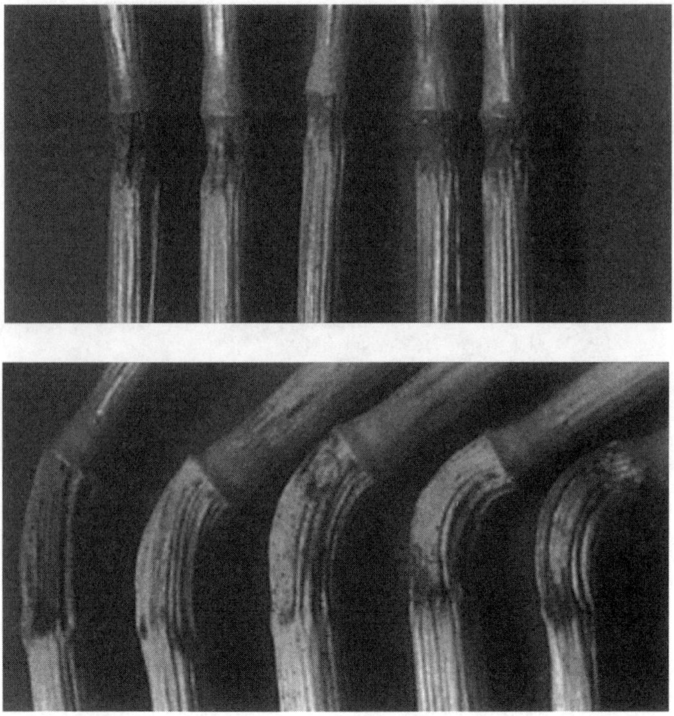

Blue Ball, Maryland, U.S. 1995. Top: Normal wheat stems and growth nodes. Bottom: Nodes lengthened 130 to 200 percent and reoriented. Photograph © 1995 by W. C. Levengood, Ph.D., biophysicist.

Swollen and reoriented growth nodes were collected in 1995 from downed wheat that had no geometry near Blue Ball, Maryland. Randomly downed crop, sometimes in odd patterns that don't follow the natural syndrome of lodging where plants fall over in very wet ground has repeatedly been found in the United States. The Blue Ball, Maryland, plants averaged 130 to 200 percent node lengthening and dramatic reorientation from the normal straight vertical. The cause is thought to be rapid heating and expansion by microwave energies, which collapse the cell tissue in the upper nodes and produce the angled reorientation.

Nodes on wheat stems closer to the ground have tougher, more mature tissue. Dr. Levengood thinks that water-cell components in these lower growth nodes build up pressure and, like safety valves, literally burst open to create what he calls "expulsion cavities." Sixty-five to 85 percent of the Blue Ball, Maryland, wheat formation plants contained expulsion cavities, while none were found in the controls. The following photographs show an excellent contrast between four exploded nodes from a wheat formation in Logan, Utah, in August 1996 compared to normal control plants in the lower image.

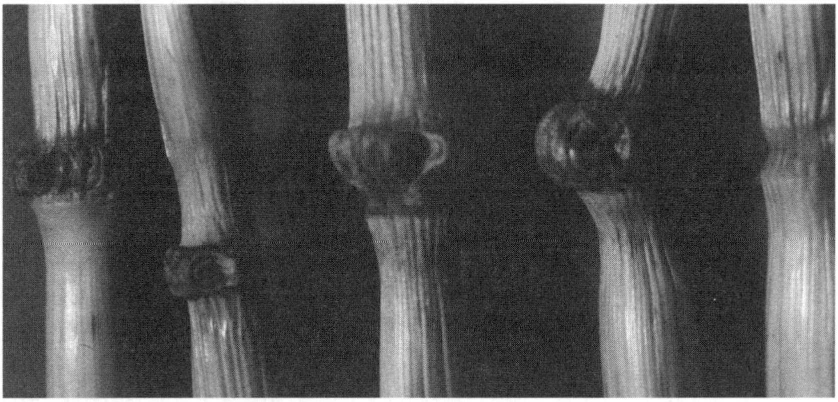

Growth node expulsion cavities from Logan, Utah, wheat formation, August 1996. Dr. Levengood thinks microwave energies heat up node cell water to explosive force. Photograph © 1996 by W. C. Levengood.

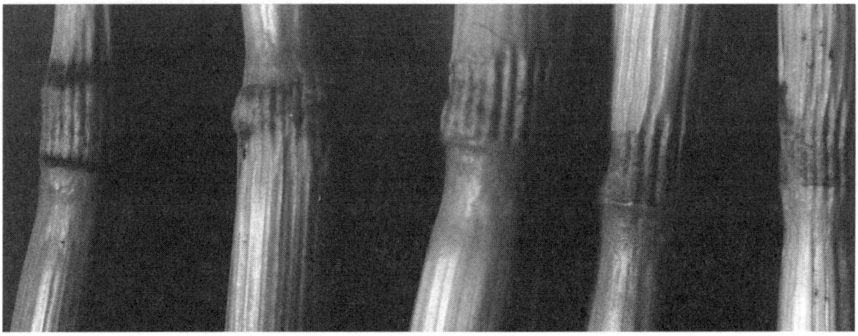

Normal growth nodes in wheat plants collected as controls in Logan, Utah, August 1996. Photograph © 1996 by W. C. Levengood.

"There is no way that wind or air motion from a helicopter or pressure from boards, feet or string could produce these biochemical and biophysical changes," Dr. Levengood said. "The plants have been affected by an intense energy."

In addition to these fundamental biophysical and biochemical changes in the wheat plants themselves, Dr. Levengood has studied plants from England and the United States that have been coated with a deposit of iron oxide, hematite and magnetite that he believes are fused particles of meteoritic origin. Peter Sorensen found the first coated wheat and soil in 1993 in a field near the 1993 formation and called me to ask what it might be. I called Dr. Levengood, who asked if Peter would drag the field with a magnet. Peter did and sent what he collected to Dr. Levengood. I went to his lab and watched him demonstrate how the coated wheat bract tissue was attracted to a magnet.

Dr. Levengood hypothesizes that the source of the iron deposits are meteoritic particles from the earth's upper atmosphere. Over 70 percent of approximately forty crop formations currently investigated with magnet drags turn up iron deposits on the surface of the soil.

Magnet attracted wheat bract tissue covered with iron presumed to be from a concentration of micrometeoritic particles. Found in field near 1993 Cherhill formation after harvest by Peter Sorensen. Photograph © 1993 by author.

While I was walking through the July 29, 1999, Avebury pattern, another visitor arrived and reported that a brand new formation had been found early that morning in a field of oats not more than a mile away. Getting into formations as soon as possible before public trampling is always a challenge, so I hurried back to the car.

When I drove past the six-cube formation and reached the top of the next hill in the road, I could see on the left a darkened area like a shadow on a beige patch of crop growing on the hillside. The way to get up the hill without trampling through the oats was to walk along a dark green hedgerow that bordered the field. I was looking for a stile in the barbed wire. Those wooden steps would get me up one side of the sharp wire and down the other. But I couldn't find one and decided to climb to the top of the hill beneath a large grove of trees and see if I could get to the formation from there. I wanted to enter as near the formation as I could to reduce walking in the growing oats.

At the top of the hill, the crop was still fenced off by barbed wire. It was so hot that I didn't have a jacket with me to put over the sharp barbs. I tried pulling open a space between the middle wires, but they were too tight. Without someone to help, I risked tearing my clothes and skin if I tried to climb, straddle and swing off the top wire. Then I heard the swish of grass in the trees behind me and two women appeared. They had a windbreaker. We covered the sharp wire, the first person over received our cameras and we followed.

Stinging nettles and thistles stung and scratched my bare legs so much that it was a relief to finally move into the oats without knowing where we were going. The formation, it turned out, was much farther below us on the hillside than it appeared from the road.

At first I stood at the edge trying to understand what the pattern was and what the plants looked like without stepping on them. I prefer to first view the plants through the zoom lens of my camera, looking for swollen nodes, burst nodes, reoriented nodes, single standing stalks or other anomalies. The two women went directly to the center and stood with their arms outstretched and eyes closed in a kind of standing meditation. Centers are often damaged or completely destroyed early on by people who feel compelled to stand, sit or lie down on them.

From my position at the outside edge of the downed oats, my first impression was how dishevelled the oats looked on the ground compared to the satin sheen that wheat can have, as if laid down by rushing water, often in opposite directions, crisscrossing patterns, ninety-degree angles and weavings like braided hair. The oats were rougher.

When I was able to look at the crumpled center, there were six oat paths coming from the outer perimeter so that seed heads merged at a central point like spokes around the hub of a wheel.

Half a dozen people were moving around in the formation, which was about 60 feet in diameter. Several men spoke German. One had a blue and white bandanna tied over his head. He was sketching on a notepad and I moved over to see his drawing. It was a six-petalled flower with numbers indicating lengths. He was Andreas Mueller, a college student from Saarbruecken, Germany, who produces the I.C.C.A.—The International Crop Circle Archive and has devoted several summers in England to surveying crop formations and producing elegant survey diagrams. (See Chapter 9.) He spoke English and I asked him what he thought of the formation around us. Andreas said he thought the lay was more complicated and organized than it might first appear, and he did not think it was hoaxed.

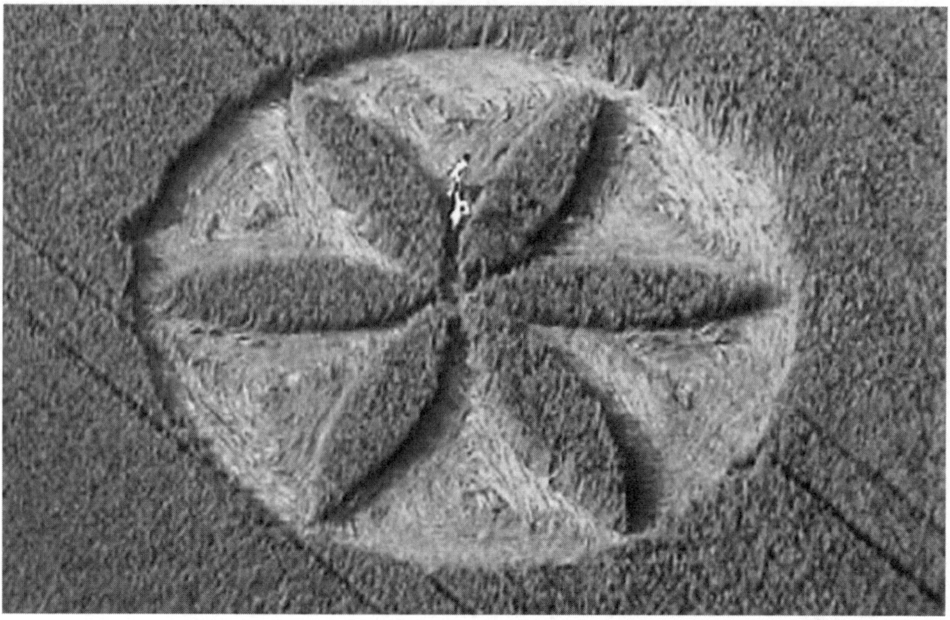

Aerial photograph of six-petalled formation in oats near Avebury, discovered Saturday, July 31, 1999, the same morning that the enormous formation at Roundway appeared. 19.2 meters (63 feet) in diameter. Videograph © 1999 by Peter R. Sorensen.

The Six-Petalled Flower is well known in sacred geometry as a demonstration that "sixness and oneness are closely related," as Miranda Lundy illustrated in her 1998 book, *Sacred Geometry.* She writes: "Sacred Geometry charts the unfolding of number in space. The basic journey is from, and eventually back to, the single point, into the line, out to the plane, through to the third dimension and so on... The circle, the triangle and the square are often thought to represent heaven, man and earth."

Lundy defines geometry as a universal language — perhaps *the* cosmic language — that describes the smallest to the largest in this universe at both seen and unseen levels. Underlying everything are mathematical rules of construction that define how matter and energy unfold and evolve. For example, a natural growth progression from a single circle is to divide its perimeter into six equal parts that become the centers of six more circles.

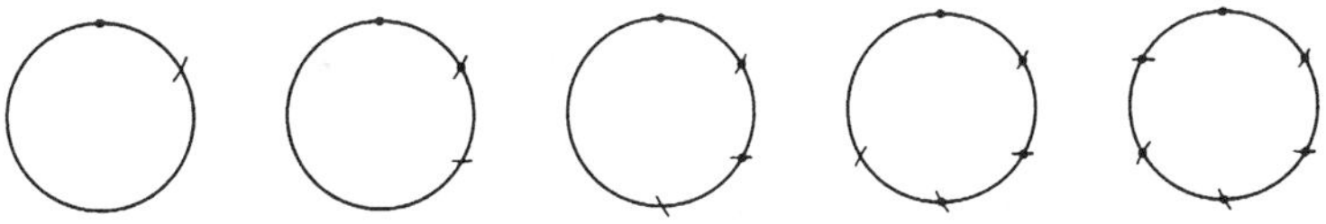

Another writer, Nigel Pennick, said in his 1994 book *Sacred Geometry, Symbolism and Purpose in Religious Structures,* "Because geometry is an image of the structure of the cosmos, it can readily be used as a symbolic system for understanding various features of the universe.…Since the earliest times, geometry has been inseparable from magic. Even the most archaic rock-scribings are geometrical in form. These hint at a notational and invocational system practiced by some ancient priesthood. Because the complexities and abstract truths expressed by geometrical forms could only be explained as reflections of the innermost truths of the world, they were held to be sacred mysteries of the highest order and were shielded from the eyes of the profane (general public).

"...Complex concepts could be transmitted from one initiate to another by means of individual geometrical symbols or combinations of them without the ignorant even realizing that any communication had taken place....Each geometrical form is invested with psychological and symbolic meaning."

Symbols and their mathematical and spiritual meanings and uses in magic rituals are part of the Mystery Schools of the Rosicrucians, Egyptian and Greek Hermetics, Kabbalists, Knight Templars, Freemasons and Mesopotamians. Magic's relationship to religion might be that magic tries to manipulate forces in the unseen to produce a certain end *beyond* the knowledge of normal humans. Religious rites are ways for ordinary humans to act out beliefs about the relationship between humankind and a Creator Force that is personal, conscious and an omnipotent spiritual being. The earliest known recorded texts of magic spells and alchemical symbols and formulas were in ancient Mesopotamia and Egypt.

The British Isles were long ago appraised by an intelligence with a profound understanding of geometries and the electromagnetic grid of our planet. Stonehenge, constructed on several geometries to keep track of cosmic cycles, was also "sited at a key geomagnetic point with regard to the overall landscape geometry of southern England," said Pennick. The great henge is sited on several important ley-lines, including one that runs from St. Michael's on top of the Glastonbury Tor all the way to a tumulus in Surrey. Stonehenge itself "shows a tendency towards a sixfold division (in the) discovery of an equilateral triangle with six-mile sides that links the microcosmic sixfold sacred geometry with macrocosmic landscape geometry." Stonehenge's ground plan even seems to be organized around an invisible hexagram. (See page 93.)

If Pennick is right, then which ancient intelligence understood the magnetic grid of the earth and its relationship to geometries on the planet and in the cosmos?

While Andreas and I were talking, another dark-haired man approached and told me he had seen orange lights over the trees at the top of the hill the night before. These lights, he said, were like the other ones he and his wife had watched at Silbury Hill the evening of August 28. That was the night of the big Avebury formation with the six 3-D cubes in it. He had even photographed one of the orange lights.

Evening of July 28, 1999. Camera is pointed from the "ziggurat" formation across the road from Silbury Hill toward Avebury, where the six-petalled pattern was found early on July 31. Photograph © 1999 by Charles Mallett.

I told him I was hoping to write a book about the link between mysterious lights and crop formations and wondered if he would talk to me on audiotape about his experiences.

"I think that would be all right, but we should go and get my wife, Frances. She's in the car on the road. She can't take the heat. My name's Charles. Charles Mallett," he said, extending his hand.

We retraced the steps up the hill through the oats, over the barbed wire, down along the hedgerow about a half mile to the car on the road. On the way, Charles asked me if I had been in the other new formation that morning at Roundway.

"Another formation this morning?"

"Yes. This one here in oats and the Roundway one in wheat were both found this morning. Roundway is one of the best ones I've seen so far. We were in the formation very early this morning. The quality and sheer scale of the thing is incredible. And the precision impossible."

"Could we go there so I can see it and also talk to you and Frances on tape?"

"Let's see what she says."

By the time we reached Frances in the car, my clothes were wet and sticking to my perspiring skin. Frances said she would be happy to do an interview, but it would have to be done in shade. So we set off on the road to Bishops Cannings. I followed Charles up an increasingly steep hill on a very narrow road. We came out on top near a forest of dark green trees. The hill sloped down into a huge, golden-beige wheat field. In the middle was a shockingly perfect pattern that measured 361 feet in diameter, bigger than a football field.

Charles and Frances Mallett on the hill above the Roundway formation in wheat on Saturday afternoon, July 31, 1999. Photograph © 1999 by author.

Roundway, Wiltshire, formation in wheat discovered at 6 AM on July 31, 1999, 110 meters (361 feet) in diameter. Aerial photograph © 1999 by Lucy Pringle.

Author Linda Moulton Howe at Roundway, Wiltshire, England, formation on Saturday, July 31, 1999, with Charles and Frances Mallett. Photograph by Charles Mallett.

Later I studied a similar pattern in Manly P. Hall's 1928 book in a chapter entitled "Fifteen Rosicrucian and Qabbalistic Diagrams." Seven circles on arms radiating from a central circular area are described as "seven spirits before the throne." The triangle core of the diagram "marks the center of eternity" and its three corners are labeled with the Hebrew words for Father, Son and Spirit. The lower concentric rings are the "angelic world ending in the cognizable world of the Sons of God."

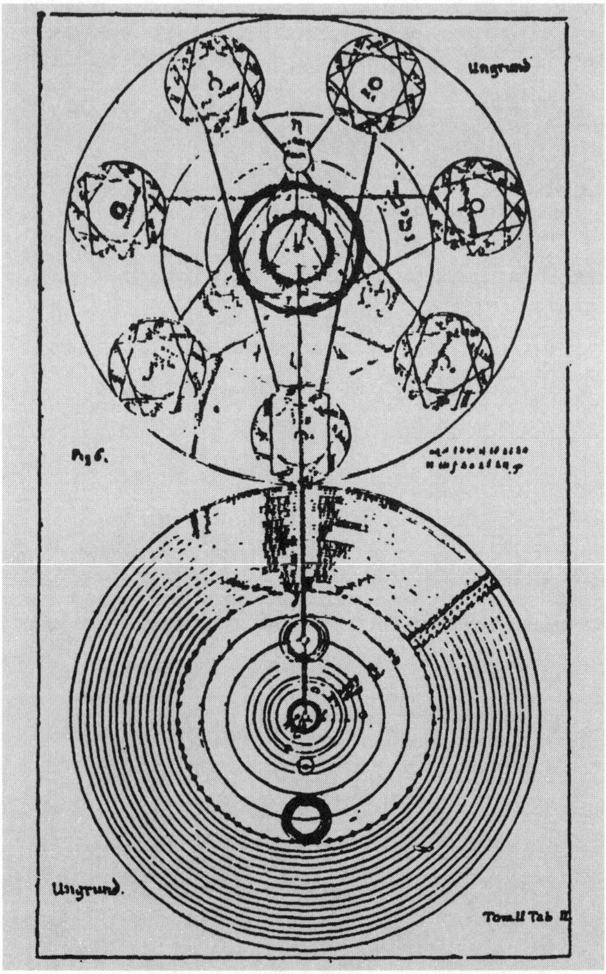

Rosicrucian and Qabbalistic Diagram, The Secret Teachings of All Ages: An Encyclopedic Outline of Masonic, Hermetic, Qabbalistic and Rosicrucian Symbolical Philosophy, *page 147.*
© 1928 by Manly Palmer Hall.

The Bible's revelation to John, Chapter 1, Verse 4, begins, "May you have grace and peace from God who is, and was, and is to come! and from the seven spirits before his throne." A throne is defined by Webster's dictionary as "the chair occupied by a sovereign; a personage who occupies a throne; the power of such a personage; and the third of the nine orders of angels."

Back in 1992, Peter Sorensen also described a connection between the Hall diagram and the great triangle at Barbury Castle found on July 17, 1991. (See page 84.) German publisher Michael Hesemann discussed Barbury Castle and the diagram and other symbols in his 1993 book *Botschaft aus dem Kosmos,* translated to English in a 1996 edition, *The Cosmic Connection.*

In the trees on the hill above the precise and elegant Roundway formation, we found cool shade. I set up my audio recorder and the three of us talked about mysterious lights and crop formations.

"A ball of light came right in front of me in a crop circle," Charles said quietly. "That was pretty weird. I decided to spend the night in a large formation which was in a field of oilseed rape. The formation was spread out right below the Alton Barnes White Horse on the side of Milk Hill. Frances and a friend of ours decided that they would spend that night on the White Horse where they could view the whole landscape."

Oilseed rape, also known as canola, is an early spring crop in England. The stems are notorious for being full of water, similar to celery stalks. No person can move through the plants without leaving a green trail. The yellow pollen and blue-white musk bloom get all over one's clothes and the thickly entangled branches and flowers make any movement through them difficult.

"When did that rape seed formation arrive?" I asked.

Frances picked up her large three-ring binder of crop circle photographs and drawings. "I've got a photo of it. Early May, I think. There were two in that field. First, there was a series of five circles. Charlie was in the second formation that appeared next to the circles."

The formation in which Charles Mallett had the light encounter appeared at Milk Hill near Alton Barnes on May 2, 1999. It looked like a delicate machine outlined in the dense, yellow flowers and, according to an eyewitness, was *not* there on the morning of May 2. Helen Robinson, a member of the Wiltshire Crop Circle Research Group, had visited the beautiful, yellow field that morning where the month-old circles from April 4 were still clearly visible. Helen left for lunch and returned in the afternoon. She climbed to the White Horse overlooking the field. To her "utter amazement" there was the brand new formation next to the old circles. That would mean the "machine" was created in clear daylight. (See aerial photos of both formations on page 106.)

Charles said, "I planned to sleep in the end ring, which had a large standing tuft in the center. Nothing in particular happened. It gets dark around half past ten. It's clear, nice night, dark. And then suddenly, it felt quite surreal. I don't know why."

"What do you mean, 'surreal'?"

"It's kind of like being detached. It's hard to explain. It's so internal. Just slightly out of sync. I could hear the roads about a quarter of a mile out in front of me. I can hear the cars and animals all around, but it all feels slightly away, slightly detached in some way."

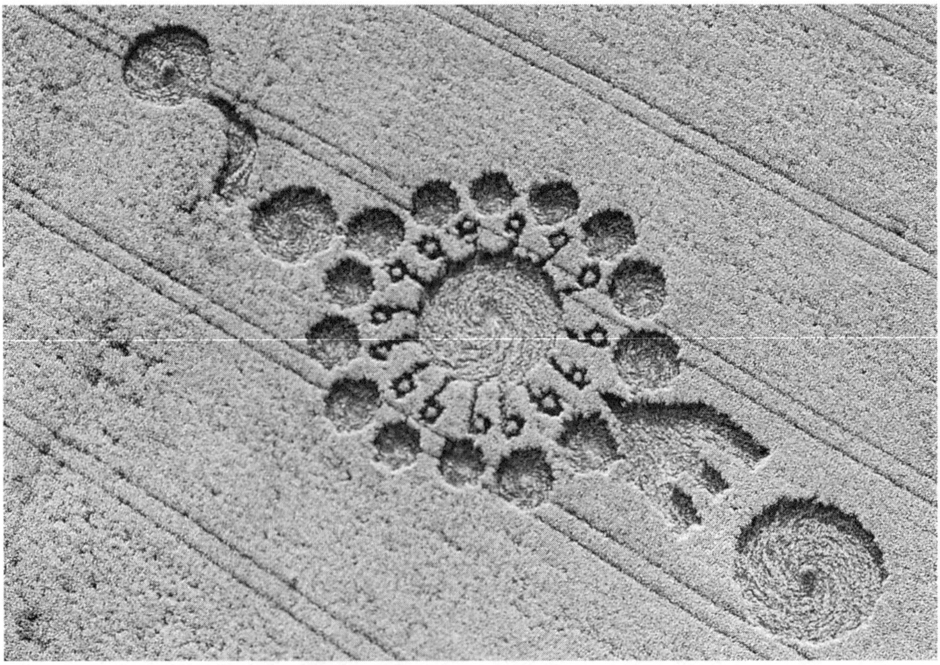

Milk Hill, Wiltshire, discovered May 2, 1999, in yellow flowering oilseed rape, 250 feet long. Charles Mallett encountered a sphere of light inside the circle at the far right. Photograph © 1999 by Steve Alexander.

"I sit there for a couple of hours, up until around midnight or 12:30 AM. I take a look up to Sirius, the brightest star in the sky. I look down and all of a sudden there's a bright, diffused luminosity. It's a sphere about ten to fifteen meters in front of me. It's just above ground level behind the standing center, which is about 2 feet across with thick stalks, and I can see between them. The stalks are in silhouette. This light is behind them and next to the back wall of the circle. I sat and watched this fully up to a minute and a half. I had my camera next to me and it didn't even occur to me to pick it up. I should have."

"Did the light move at all?"

"There was hardly any motion. Very, very slight, but really no motion. And it was yellowish-white, diffused with a hint of orange, and appeared to be self-luminous. But it wasn't casting a bright light at all. It was glowing, but very localized. I had enough time to shift my position, move away from the standing center so I could look around, and it was definitely there. I'm actually sleeping in these crop circles, hoping for some kind of direct interaction with whatever is doing this stuff, and my mind goes blank. All I seemed able to do was watch and observe what was going on right in front of me. After a minute and a half or whatever — long enough to be absolutely sure that this object is there — it fades out.

Yellowish-white glowing sphere Charles Mallett saw suddenly appear behind oilseed rape in Milk Hill formation, 1999. The light stayed for more than a minute. Drawing © 1999 for author by Charles Mallett.

"I must say that sitting inside a rape seed crop circle, the plants are about five feet high. There is absolutely no way whatsoever that another person could have been in there. The tramlines are all overgrown. It's like a jungle. You can hear someone coming one hundred meters away easily. There was no one else in there. I can't account for the light then or now. Very weird!"

"How exactly did the light come and go?"

"When I first saw it, I had been looking at Sirius below the Orion's Belt constellation. I looked down and it was just there. I didn't see it pop in or fade in or anything like that at all. Then, after a minute or so, it just faded out over about two seconds to nothing."

"Did you ever feel like the light tried to communicate to your mind?"

"No, no, I'm pretty sure I was in control of my mind. But I had rather strange, almost spooky sensations for about fifteen or twenty minutes before I saw anything. That was quite weird, a heady, slightly phased-out-of-reality feeling, even before I saw the light. I considered leaving the crop circle because it was so strange. Yet I'm used to sleeping in crop circles and on the hills in this area. Even though sometimes it gets a little strange, I have never had any feelings of negativity from whatever is responsible for the formations. Quite the opposite, in fact. The crop circles seem to have a hugely positive effect on most people that come in contact with them.

"I didn't feel I was threatened when this light was there. As soon as the light showed up, I was transfixed and quite excited. It's right in front of me, what I've been looking for. I think it was like if you were in an enclosed space and you can hear things going on around you outside, it's slightly worrying. But it's not nearly as worrying when you can actually see what's going on. The feeling of a presence is quite strange. But if you can see, it's not nearly as strange."

Frances added, "The same night in May [1999] that Charlie saw his sphere, I was at the belly of the White Horse at Alton Barnes with a friend of mine, and we were staying the night overlooking the field that Charlie was in.

"About 10:30 PM or maybe a bit later, two lights appear very bright in the field where Charlie was in the rape seed formation. And if you've ever been into rape, it is so difficult. You make so much noise. You can crunch the stems so easily. But that night right above the rape, these two or three lights flashed on, went off, flashed on, went off. And that was it. Until about a half hour later, I saw these streaks of light — a bit like smoke.

They didn't have a source. They were just streaking down about the same place as the formation. The streaks were almost working their way down from the top of the formation to the bottom.

"I didn't know that Charlie was having the light experience in the rape while we were watching above the rape until we went to meet him in the formation the following morning. He said, 'You will not believe what I saw in here last night!' We said, 'We saw some lights in this field, too!'"

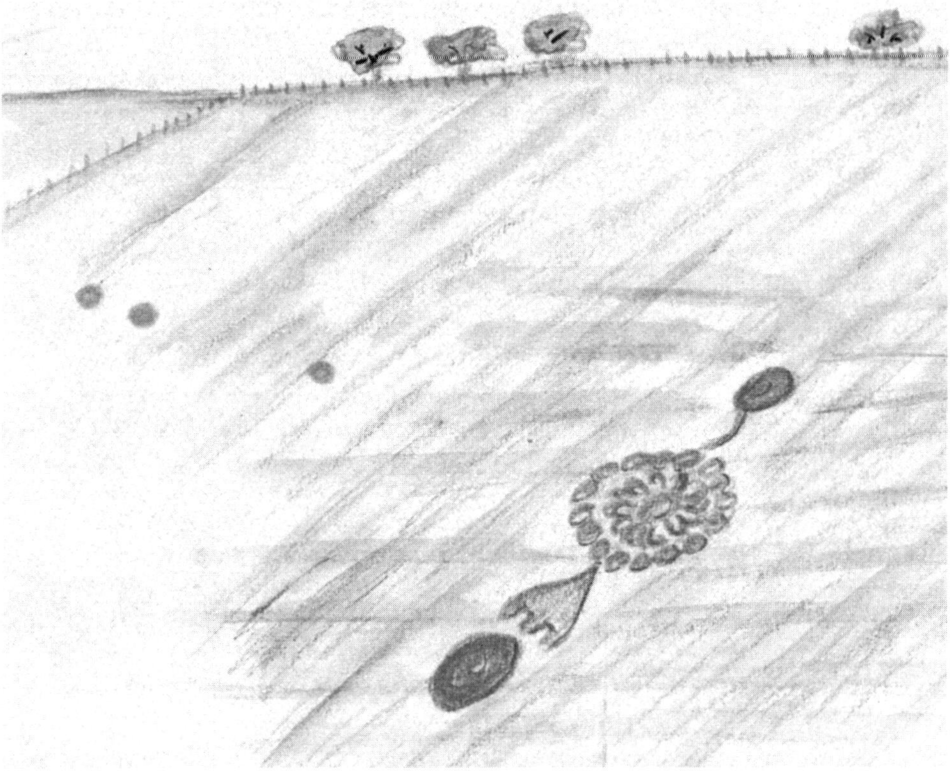

Three mysterious lights Frances Mallett and her friend saw to the left of the Milk Hill, Wiltshire, oilseed rape formation the same night in May 1999 that Charles Mallett encountered a sphere of light inside this formation. Drawing © 1999 by Frances Mallett.

Charles Mallett told me he works as a care officer in the region of Horton, Wiltshire, where he lives with Frances and her two children from a previous relationship. Charles goes to the homes of elderly and disabled people to help with shopping and housekeeping. He was born in the 1970s in Devon, where he dropped out of school at sixteen. He said, "I have been on a very interesting and sometimes very deep search in the last few years for some truth about many aspects of life and reality. Crop circles seemed to act as a catalyst for my search."

Frances was born in Lockeridge not far from Alton Barnes in the heart of Wiltshire crop circle country. She studied interior design in college and likes to paint the formations and their geometries in watercolor. She also works for a nonprofit charity trying to help alleviate the poverty in northern Albania and Kosovo.

They married in 1998 and have lived in Horton between Roundway and Alton Barnes, where they are close to the crop circles each summer.

Charles and Frances Mallett in the garden at the Barge Inn, Alton Barnes, Wiltshire, England, August 1, 1999. Photograph © 1999 by author.

Frances flipped to another page in her ring binder. There was a beautiful gold "ziggurat" pattern seeming to rise three-dimensionally from the wheat. Ziggurats were terraced pyramid temples first built by the ancient Sumerians in Mesopotamia (modern-day Iraq). The ziggurats were named "Mountain of God" or "Hill of Heaven." The temples ascended to a small top chamber where a god or gods could dwell and sleep. Selected human females would also periodically climb the long stairs to the room on top.

The 1999 formation had a repeating fractal pattern of ninety-six small circles in triangular groups of three and sixteen larger circles along the twelve triangular edges.

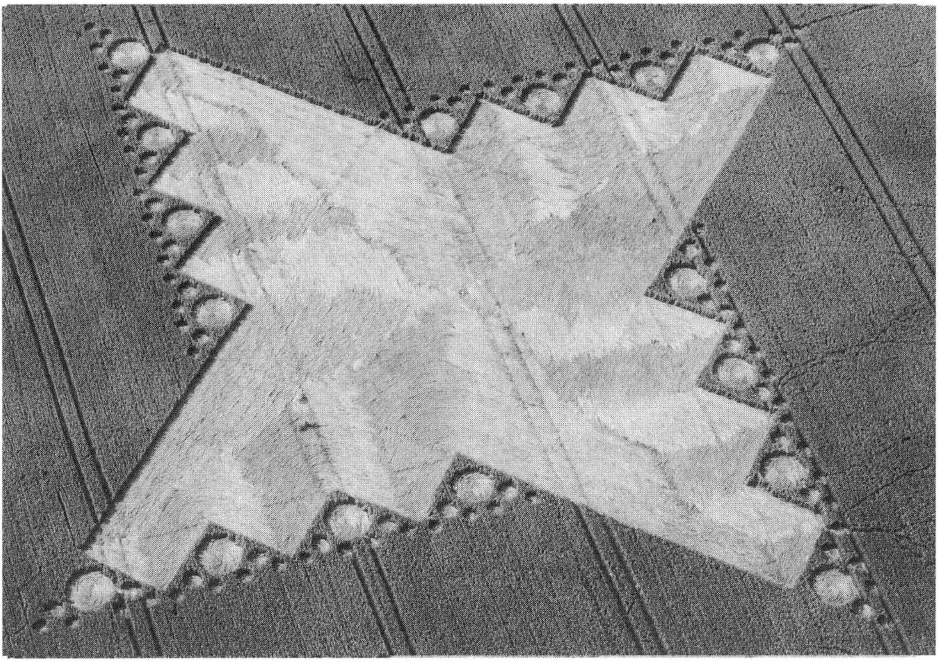

"Ziggurat" formation was discovered the morning of July 24, 1999, in wheat across the road from Silbury Hill, longest axis 97.5 meters (320 feet). Photograph © 1999 by Steve Alexander.

On the night of July 23–24, a man named Mike Ray stayed all night in the Silbury Hill car park across the street from the wheat field. He told researcher Stuart Dike of cropcircleconnector.com that before sunup on July 24, he saw the field glowing, but could not see a light source.

Later in the afternoon of July 24, 1999, Joachim Koch, M.D., and a colleague, Hans-Juergen Kyborg, were on Knap Hill. Dr. Koch had organized a session for about thirty people to gather that night to meditate on crop formations and world healing on top of Woodborough Hill. The warm day was beautiful, and Dr. Koch aimed his camera across the barley in the East Field at the beech trees atop Woodborough Hill where they would all be meditating in a few hours. The two men suddenly saw a bright light and photographed it.

Camera pointed from Knap Hill across barley in East Field to the beech trees on top of Woodborough Hill, July 24, 1999. The light below the trees was visible at the time. Photograph © 1999 by Joachim Koch, M.D.

Dr. Koch wrote to me about his photograph: "This extraordinary, genuine photograph is another indication of involvement by non-human intelligence in the crop circle phenomenon as it shows an unknown aerial object, perhaps from another dimension. It appeared out of nowhere as a silvery, glowing, spherical light of approximately four meters in diameter in comparison to the trees on top of Woodborough Hill. First, it hovered left of the hill for more than ten minutes. Then it started to move to the right, stopped and hovered above the trees for another ten minutes. Finally, it turned again and landed on the meadow in front of the trees where it faded and disappeared. This photograph shows the object when it already had dimmed to about half its original brightness.

"The appearance of this glowing sphere is a highlight in the experimental work that I, Joachim Koch, 50 years of age, and Hans-Juergen Kyborg, 69, both from Berlin, Germany, have been conducting in Wiltshire and Germany for more than ten years. Our experiments are based on a thorough analysis of the early pictograms from 1987 to 1990 in which we found a certain pattern of encoded astronomical information. Finally, we were able to decode this information and went to England in 1991 to prove our theory either right or wrong. We tested by giving back a message to the involved intelligence with a pictogram we made by using the same code. Amazingly, we received two answer-pictograms at Preshute Down and Barbury Castle within three days.

"The decoding of these pictograms resulted in the identification of a yellow star, HD 42807, which might harbor life on its second orbiting planet. The same star seems to be mentioned in the old Egyptian Pyramid texts and in the tomb of Senenmut in Deir al Bahari, Thebes, an enigma that we believe we have also solved.

"Since 1991, we have worked quietly in the crop circle phenomenon and believe we have established continuous communication with the hitherto unknown, benign, other-dimensional or extraterrestrial intelligence. The information we have received has helped us to create inter-dimensional links which resemble the light in the photograph."

Is this the same visible and mysterious light phenomenon recorded in videotapes by Steve Alexander, Constantin and Dominic von Durckheim, Carl Nevin, Bert Janssen, Foeke Kootje, Ed Sherwood and others? Is it connected to the "*non*-visible light balls" discussed in later chapters that have also shown up in numerous photographs after film has been developed and printed?

Four nights later, on July 28, Charles and Frances stayed out in the fractal "ziggurat" across the road from Silbury Hill.

Frances said, "I think it was 9:40 PM. We were sitting in the center of the formation looking over towards Marlborough, so we had our backs to Avebury. And a light appeared over some trees — where you were in the flower formation. You know those trees above the oats? Over there, this orange light just appeared out of nowhere and stayed quite still.

"Then we saw that it was doing very subtle movements. I said, 'Charlie, did you see that go left?' He said, 'Yes,' confirming what I could see. That light went away and in a second reappeared over West Woods, which must be about four to five miles away. Then the next second, it was back in the same place. And then another one appeared right along by its side, two of them, and both were doing very odd patterns in the sky."

Frances, Charles and so many other regular sky watchers are frustrated by chronic media explanations that unidentified aerials are simply flares, balloons or aircraft. She said, "We're quite experienced in looking at the sky. We know what's an airplane, what's a helicopter, what's a shooting star and satellite and everything else. These lights were not conventional. We were absolutely convinced we were watching a crop formation go down. The whole event went on for three hours!"

Charles added, "Helicopters turned up, the military. Very weird. They approached this site where the lights were. And as soon as the helicopters turn up, the lights blink out. Then the helicopters shoot off right past us at very high speed. I rarely see helicopters going that fast. And as soon as the helicopters take off, a light comes back on and it continues in its motions, operating in a very localized area doing very small arcs."

Frances said the lights seemed to be above a field, moving up and down. "This whole event goes on for at least a couple of hours with us just sitting there watching. The next day, we had the Avebury one — the triangle with the six cubes in the big circle."

Five thousand-year-old Silbury Hill rises 130 feet high across the road from the "ziggurat" wheat formation. Photograph © 1999 by Steve Alexander.

"And that was behind you?"

"Right behind us!" Frances exclaimed with frustration. "We just did not consider looking behind us. And that field was only about a half mile away from us past Silbury Hill."

"What's the most number of lights you saw at any one time?"

"There was just the one light in the beginning," Frances explained. "That went on for a short while, and then there were just two lights. Obviously, when you're staring at an object long enough, your eyes can play tricks. Charlie thought he saw one split. I didn't see that. There were just two, but mainly there was one that was orange, really deep orange."

"Like the inside of a cantaloupe?"

"You know these orange streetlights you get around here? Very deep orange streetlights. Very similar to that orange, I would say, from my perception anyway. And arcing, very deliberate movements."

"Frances, when you say 'arcing,' what exactly do you mean? Please show me."

Frances cupped her hand palm down, put it in the air in front of her and rocked her arched hand in small motions. "It was very subtle movements, but seemed purposeful. In the morning, we went to see if there was a formation and there wasn't. Not finding a formation at dawn was a very big surprise. It left a question mark hanging about what had we really seen. What had happened?"

"And last night on July 30th," Charles said, "I slept back out there where that six-pointed flower is and saw a couple of orange lights through my binoculars. They were very bright, obvious lights that were tangibly there. I watched them merge together and turn into a bright light and go out."

Frances added, "It sounds very much the same as a friend of mine, Nikki, saw with her brother. They came up from Barnstable last year [August 1998]. The brother was completely skeptical. We took him to all sorts of formations and he just sat there bored. Both he and Nikki stayed up at Adam's Grave on the night that the Tawsmead Copse 'Magnificent Seven' formation went down. They saw lights that split into three — a red one and two white ones. It was a bright, full moon and they said they could actually see the crop go down."

The Malletts' friend, Nikki Saville, told me a couple of days later at the Barge Inn that she and her brother, Andrew, had driven up to the top of Adam's Grave around 11 PM to watch a lunar eclipse on the night of August 8, 1998.

"Then about 11:30 PM in the distance to the left of us, Andrew saw two red lights. By the time I looked, it was just two very clear white lights one directly above the other, very round, which only lasted a second, maybe. They were like two bubbles bursting, that's how it looked to me, just like they had been popped. And they disappeared.

"And then we saw one very bright light go across the landscape to almost opposite of where we were. We were paying a lot of attention to the moon, really, and the moon was shining up the ground below. And we could see this ball of light stop at what looked to be along the horizon line and then break into three that moved down over the landscape. And it kind of looks a bit like the sun, or like looking at a cell under a microscope. You know, with a dark rim around the outside. They were just very round, smooth lights, not like a star, not like dazzling.

"That first light stayed up and three came from it and moved down diagonally. They just seemed to hover, not make a great deal of movement. But then as the shadow was coming over the moon, like about halfway like a half moon, we could see indents go down in the corn. [Author's note: Actually wheat. "Corn" is a generic English term for cereal crops.] We only saw half of the formation to start from where we were sitting. It turned out to be a big circle.

Tawsmead Copse "Magnificent Seven," Wiltshire, August 9, 1998, 300 feet in diameter; 175 little circles, each with a different twisted center like a bird's nest. Survey © 1998 by Peter R. Sorensen.

“And we sat there until about half past 5 in the morning and there were lights moving around in the area the whole time. Tawsmead Copse was, I think, about a quarter of a mile from our eyes. The actual formation looked like an inch wide from where we were. We went down after it got light. We were in the formation just after 6 AM. It was very exciting. We had been in others the day before and we knew this hadn’t been there. You could tell it was so fresh, we were just speechless!”

Seven-fold geometry is rare in crop circles, but one month before the Magnificent Seven formation at Tawsmead Copse, another seven-fold geometry had appeared in the East Field a mile away. Researcher Bert Janssen said the logical geometry construction evolution would have been from Tawsmead Copse to the East Field, but the phenomenon put the patterns down in reverse order. Photograph © 1998 by Janet Ossebaard.

As Charles, Frances Mallett and I sat talking at Roundway, I asked Charles if he had ever encountered any other light spheres before.

"Not a light, but a very weird sound. The first time I came here to Wiltshire to find some answers was the beginning of the 1997 season. I think it is very important to come here and look at the formations on the ground firsthand. And ever since I've been coming up here, I've had bizarre experiences. That one earlier this year is the closest action I've ever had. But I've had strange sounds, strange feelings, quite often weird lights. But not in such close proximity to me as that rape seed experience with that sphere."

"What strange sounds have you heard?"

"One in particular was the first occasion I came to Wiltshire to look at the crop formations. I was wandering the hills looking for my first crop circle. I really wanted to look at one on the ground, which is really different than looking at pictures. I wanted to see how the crop's laid. And a real nice one turned up at Alton Priors. It was called the Torus Knot crop glyph at Alton Priors, across the valley from Adam's Grave. It was huge, a quarter mile across and in a straight line with the East Field below Adam's Grave. This year in 1999 there were those two big formations in that same field." ("Sentence" and "Serpent"; see Chapter 6.)

Crop formations have repeatedly appeared over the 1990s in crops near Adam's Grave, Alton Barnes, Woodborough Hill, Alton Priors, Milk Hill, Knap Hill, Tan Hill, Golden Ball Hill and Stanton St. Bernard. Before continuing Charles Mallett's experience with the unusual sound, it would be valuable to learn more about the heart of the crop circle activity known as the Vale of Pewsey.

Vale of Pewsey

The Vale is featured by authors Julie and John Wakefield in their 1999 book *Legendary Landscapes, Secrets of Ancient Wiltshire Revealed.* They describe the Vale of Pewsey as "part of the Marlborough downland in North Wiltshire, just three miles as the crow flies from Silbury Hill. The area is one of outstanding beauty, with its stretches of farmland and spectacular, undulating hills. Nestling in the heart of the Vale are the quaint, sleepy villages of All Cannings, Stanton St. Bernard, Alton Barnes and Alton Priors. The range of hills that are relevant to this study are those of the south-facing escarpment."

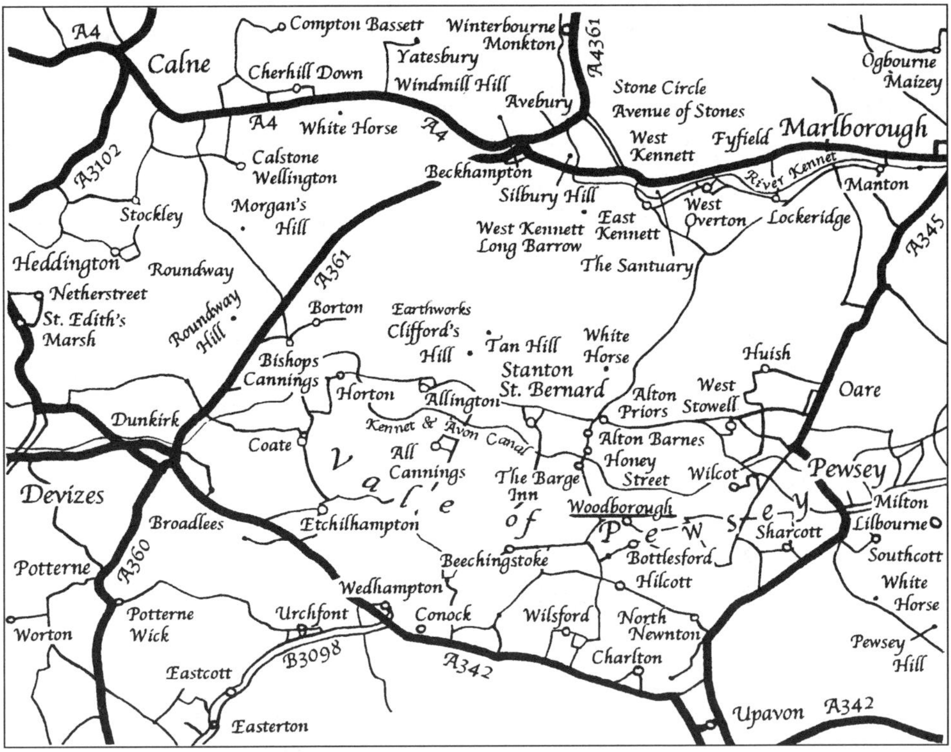

Vale of Pewsey, a large valley between Devizes and Pewsey southwest of Marlborough, through which runs the River Kennet and a series of connected hills including Tan Hill, Milk Hill, Adam's Grave, Knap Hill, Golden Ball Hill, Martinsell Hill and the villages of Etchilhampton, Stanton St. Bernard, Alton Barnes, Alton Priors, Wilcot, West Stowell and Pewsey. At the heart of the Vale is Woodborough Hill. Map © 2000 by author.

The south-facing hills include Tan Hill and Milk Hill, which are the highest points in Wiltshire at about 900 feet. The Wakefields write, "There is a legend that the body of Queen Guinevere and her funeral cortege traveled over Tan Hill at night on their way to her interment at Glastonbury where her husband King Arthur was buried. On the pinnacle of Milk Hill is a barrow and a white horse was carved into the side of the hill in 1812.

"The adjacent Walker's Hill is capped by a Neolithic chambered barrow known as Adam's Grave, the most dramatically sited long barrow in Wiltshire and possibly in Wessex. According to local folklore, a giant emerges from Adam's Grave if a person runs around it seven times. The prehistoric trackway known as the Ridgeway crosses the col [pass] between Walker's Hill and Knap Hill, along which sarsen stones were brought from the Marlborough Downs to build Stonehenge." Other scholars suggest the huge stones were rafted down the River Avon.

"In Druid mythology, one of the names for the sun was 'A-dda,' and the Christian monks associated this with *Adam,* the Hebrew word for 'Man.' In rabbinic tradition, Eve's name, *Havvah,* is associated with the Aramaic *hewyah* which means 'serpent.'"

Serpent motifs have repeated in the crop formations throughout the 1990s. Two serpents were laid down in crop formations in 1999: one in the East Field at Alton Barnes on June 12 (See Chapter 6.) The other was discovered near Barbury Castle north of Avebury on August 5. I got to that one on Friday, August 6, and walked from the elegant arc above the large circle "head" all the way to its beautiful coiled and layered tail.

One end of upper arc over "Serpent's head," where wheat laid forward from a narrow point. Photograph © August 6, 1999, by author.

"Serpent" formation discovered August 5, 1999, at Barbury Castle, the area's fourth in 1999 after the Menorah, "scarab" and triple crescents. Videograph © 1999 by Peter R. Sorensen.

I counted off steps inside its long, sinuous path at close to 800 feet. The path's width was about 5 feet. At the beautiful spiraled "tail," the wheat narrowed down ending with a loose weave of four bunches of wheat stems overlapping each other in ninety-degree angles.

In a 1937 book, *The Druids' Temple Near Glasgow* (Scotland), writer MacLellan Mann investigated serpent symbols in Druidic ancient astronomy. He found that various serpent symbols, differentiated by length, were related to specific heavenly bodies such as the sun, moon, planets and eclipses.

The Druids carved rocks with rings and depressions that were later found to have a relationship with star patterns and sun and moon eclipses, including the 18-year, 11-day Saros Cycle related to repeated cycles of total solar and lunar eclipses on the earth over a period of nearly a thousand years. (See Chapter 4.)

End of "Serpent's tail" at formation near Barbury Castle, August 6, 1999. Photograph © 1999 by author.

Another serpent reference that connects Wiltshire to Egypt is the Wakefields' discovery that the word "Knap" in Knap Hill can be traced back to the Egyptian word *Cneph,* which derives from the Hebrew *ganaph volare,* meaning to fly, or a wing. The Wakefields discovered that Cneph was the name for the winged Egyptian deity that held a serpent and a circle in his hand.

They found archaeological links between Wiltshire and Egypt: "Knap Hill is surmounted by a Neolithic causewayed camp established around 3500 B.C. A causewayed camp is an oval or circular enclosure built by the early farmers, usually on a hilltop, and surrounded by concentric banks and ditches. The ditches being interrupted by numerous causeways of untouched chalk. Their purpose is somewhat confusing for there is no evidence that they were permanent settlements, and they do not seem to have been used as fortresses. The term 'causewayed camp' is therefore deceiving and some writers prefer to call them causewayed enclosures.

"Seventy-seven Egyptian faience beads dating from approximately 1600 B.C. were discovered in twenty three Wiltshire barrows, including a barrow on Tan Hill, and this is eight more than have been found in the rest of Great Britain and western Europe."

In addition, "the enclosure on Knap Hill is reputed to contain treasure and it overlooks the expansive East Field. It shows no evidence of defence and Sir R. C. Hoare stated that he could 'discover no apparent signs of any extensive British population at that enclosure on Knap Hill.' Furthermore, an unusual serpentine bank and ditch that leaves the enclosure and runs down the steepest part of Knap Hill has no logical function."

The Wakefields also learned that "Knap Hill is connected to Golden Ball Hill in the east by a level saddle of triangular shaped land, onto which a triangular enclosure was built in the Late Celtic era. Archaeologists recently uncovered what appears to be the remains of a Mesolithic camp on Golden Ball Hill which indicates that it was inhabited 7000 years ago and is perhaps the oldest communal dwelling in Britain."

Adam's Grave is a chambered barrow 200 feet long and 20 feet high that rests like a nipple on the breast of Walker's Hill. Sir R. C. Hoare wrote, "If I were to fix on any one artificial mound within Wiltshire county likely to have been appropriated to services of worship, I should name the Long Barrow on Walker's Hill, differing from others in its construction and protected by a bank and ditch towards the north and south. Adam's Grave was excavated by J. Thurnam in 1860 who discovered the remains of three, possibly four, skeletons within the ruined burial chamber at the east end. Found near the skeletal material was a leaf-shaped flint arrowhead. A similar arrowhead was recovered from West Kennett Long Barrow. Arrowheads were believed to have protective powers."

The Wakefields learned, "During his excavations Thurnam also found evidence of dry-stone walling in Adam's Grave and at West Kennett Long Barrow. The spaces between some of the sarsen uprights that formed the walls were filled in with panels of the same walling — oolithic limestone slabs brought from the Bath-Frome region, twenty miles to the west. This limestone was also discovered in the Occupation Area at the West Kennett Avenue in Avebury and in a pit inside a square enclosure on Windmill Hill near Avebury that is believed to have been used for the excarnation (flesh removal) of corpses. This indicates that the stone may have had a magico-religious significance connected with the dead."

During research for my two-volume book *Glimpses of Other Realities*, I discovered that Sir Arthur Keith, in his book *Al-Ubaid*, wrote, "The Neolithic people of English long barrows were also related to the Sumerians — perhaps distantly; the Sumerian type made its appearance in Europe in Paleolithic times, for one of the earliest of Aurignacian skulls found at Combe Capelle in the Dordogne, France is near akin to the ancient Arab type."

The Sumerians built massive, multilevel ziggurats (step pyramids). One might have been the Tower of Babel referenced in the Bible. Herodotus, the famous Greek historian of the fifth century B.C., described the shrine on top of Babylon's ziggurat built for the god Marduk and said it contained a great golden couch on which a woman spent the night alone, suggesting perhaps a sexual liaison with the god — and blue-blooded sons of gods?

Archaeologist C. Leonard Wooley wrote about the unusual relationship between the Sumerian human community and their gods in his 1965 classic *The Sumerians:* "While terrifyingly aloof, the gods were at the same time particularly close to man. The religion was anthropomorphic [human characteristics were ascribed to the gods] and the gods were but men writ large. The [ziggurat] temples were their houses in the city's midst...[and the gods were] eating the meats of sacrifice of which their worshippers also partook.

"The fact is that throughout the religion of the Sumerians, [it] is one not of love but of fear, fear whose limits are confined to this present life, fear of Beings all-powerful, capricious, unmoral."

Ancient Sumerian scholar and author Zecharia Sitchin has concluded from his translation work that the Sumerian "gods" were an advanced intelligence that eons ago used genetic manipulation in already-evolving primates to create "primitive workers" to mine gold and do other physical labors on this planet. Later, through subsequent genetic manipulations of primitive *Homo sapiens* by these "gods," the Sumerian human culture evolved under the stern watch of these intimidating and manipulating gods.

In 1983 at Kirtland Air Force Base in Albuquerque, New Mexico, while developing a TV documentary for Home Box Office, I was shown an alleged briefing paper for the president of the United States about extraterrestrial biological entities that specifically stated: "These extraterrestrials manipulated DNA in already-evolving primates to create *Homo sapiens*."

A Very Strange Sound

Back in July 1997, Charles Mallett spent all night below Adam's Grave in the Alton Priors wheat formation he called the "Torus Knot."

Charles said, "The formation was a central flattened circle with twelve rings. It looked like a spirograph, something like a web or knot and very three-dimensional looking.

"Torus Knot," 300 feet in diameter, discovered early on July 11, 1997, in Alton Priors, Wiltshire, across from the Alton Barnes East Field. Aerial photograph © 1999 by Lucy Pringle.

"Around about midnight, I hear a very distant sound, very high directly above me, what I assumed to be thunder or maybe military shelling on the Salisbury Plain, which is about twenty miles away in the distance. Over a course of a few minutes, these noises become uniform and seem to be coming down closer to me. After several minutes, I'm beginning to think, 'This is quite weird.' But I felt not at all worried.

"After a few minutes, the sounds are coming down towards me and turn into a uniform, gently phased heartbeat sound. Over and over again. Thump. Thump-thump. Thump. Thump-thump. And it's so close, I can feel the air vibrating in my face. This moves from my left all the way around to my right, 180 degrees around me, at a uniform distance like having a loudspeaker in front of you at 10 meters away or probably closer. There is nothing to be seen — just noises moving in half circles around me. Then it just stops.

"And that's it. I fall asleep and wake up the next morning thinking, 'That was very odd!'

"That's the night after I first came to Wiltshire to look at crop circles! And the Spiral Knot was my *first* crop circle. It's amazing. That formation convinced me quite quickly that people could not have made this. It was very, very fluid as if water had whipped around it. I'm a sensible person and I concluded that people couldn't possibly have done that in the dark. It was even overlapping the top of a hill! Standing in the formation, it was impossible to see more than 15 to 20 percent of the whole pattern at any one time. This to me makes the idea of a hoax impossible."

Frances added, "I can remember sitting up on top of Adam's Grave thinking, 'My God, this is beautiful!' You have the Torus just over East Field and to your right [near Tawsmead Copse] you had the most amazing snowflake formation."

Frances was looking below the White Horse at Milk Hill at a stunning double Von Koch Curve, or snowflake, which is a mathematical fractal figure. One was superimposed on the center of another larger one. The double Koch Curves appeared on August 8, 1997. This one broke all records, with 144 straight edges and 198 small circles. The diameter was 200 feet in wheat and very similar to an earlier single Koch Curve at Silbury Hill.

The Koch Curve is a fractal design that starts with a triangle. On the middle third of each side, a smaller triangle one third the size of the original is placed. The first result is a Star of David hexagram. When the pattern is repeated, it goes on creating an infinitely long line of triangles that surrounds a finite area. If you drew a circle around the original triangle, the Koch Curve would never extend beyond it. But the border becomes increasingly complex, perhaps an analogy for how the finite matter worlds increase in number and complexity inside the infinite force of creation.

Milk Hill White Horse in Wiltshire above a double Von Koch Snowflake fractal that was discovered in wheat on August 8, 1997, 264 feet in diameter with 198 small circles. Aerial photograph © 1997 by Lucy Pringle.

The Koch Curve is a continuous loop never intersecting itself because the new triangles on each side are always small enough to avoid bumping into each other. The Koch Curve squeezes a line of infinite length into a small area. An example in biology of this same compressed architecture is lungs. Lung tissue branches and folds over and over again to absorb oxygen. If that tissue were flattened out, it would equal the size of a tennis court!

The word "fractal" is derived from the Latin verb *frangere*, meaning to break, and the English words "fracture" and "fraction." Fractal geometry reveals that the way trees grow along the edge of a lake, the way ice cracks, clouds form, or prices go up and down, has an eerie mathematical order.

Science writer James Gleick wrote in his 1989 book *Chaos:* "Above all, fractal means self-similar.... shapes are defined, not by solving an equation once, but by repeating it in a feedback loop. You take a complex number, multiply it by itself, and add the original number and keep doing that over and over.... In the mind's eye, a fractal is a way of seeing infinity."

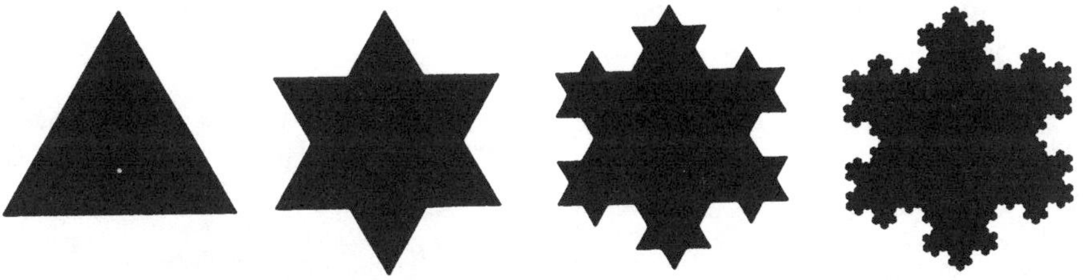

Creation of fractal known as Von Koch's Snowflake Curve.
Diagram © by Benoit Mandelbrot.

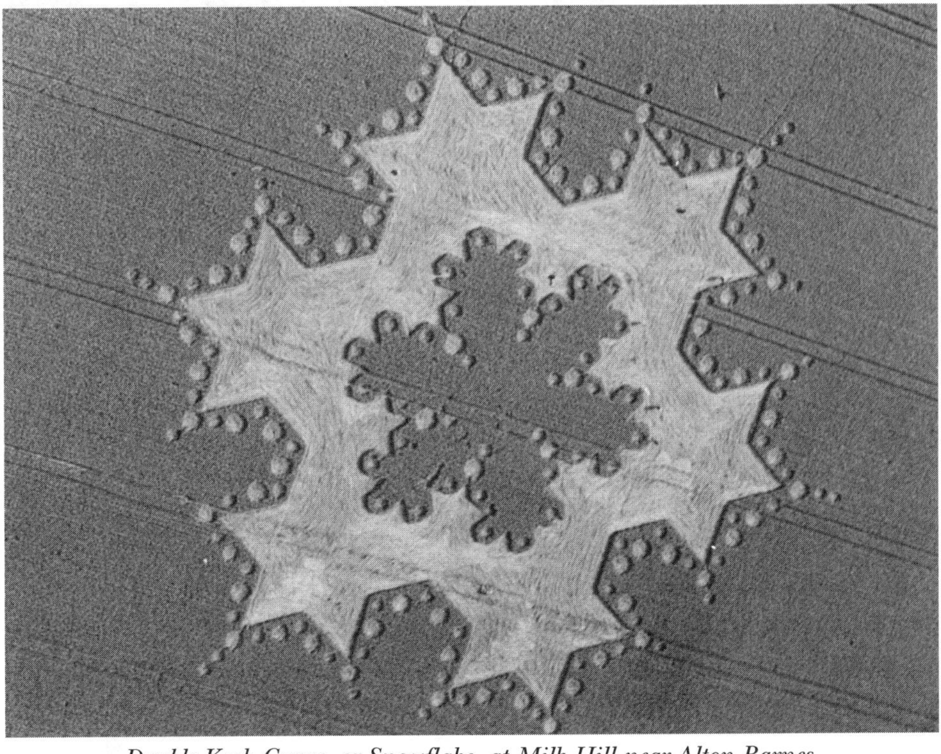

Double Koch Curve, or Snowflake, at Milk Hill near Alton Barnes, Wiltshire, discovered August 8, 1997, in wheat, 264 feet in diameter with 198 small circles. Aerial photograph © 1997 by Lucy Pringle.

A year earlier, on July 8, 1996, across the road from the huge ring of Stonehenge, a spiral arm consisting of 149 circles that covered a square acre appeared in a wheat field.

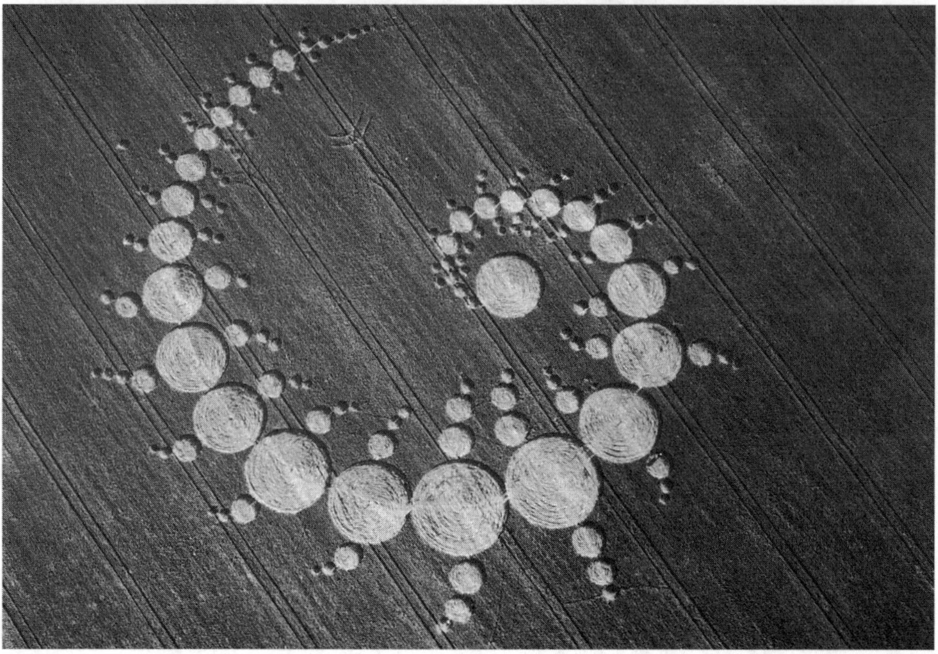

The fractal "Julia Set" discovered in a wheat field across the road from Stonehenge on July 8, 1996. The spiral measured 915 feet on its longest axis, covered a square acre and consisted of 149 circles. Photographer Peter Sorensen walked its entire length and said every circle had a different twisted center. Photograph © 1996 by Steve Alexander.

Three witnesses — a pilot, a groundskeeper and a security man — told investigators that the field was normal at 5:30 PM. About forty-five minutes later, the pilot came back past the field and discovered the circles. The spiral design matched a mathematically produced image called a Julia Set first discovered during World War I by the French mathematicians Gaston Julia and Pierre Fatou.

Later, in the 1960s and 1970s, IBM researcher Benoit Mandelbrot explored the repetition of complicated processes and found that he could create one image that would serve as a catalogue of Julia Sets. His research evolved to develop the field of fractal geometry now used to study Chaos Theory and its application to patterns of growth in the natural world.

On July 29, 1996, one of the most astonishing crop formations since the phenomenon began appeared at Windmill Hill near Avebury. A triple Julia Set spiraled a thousand feet from side to side and the number of circles was 194.

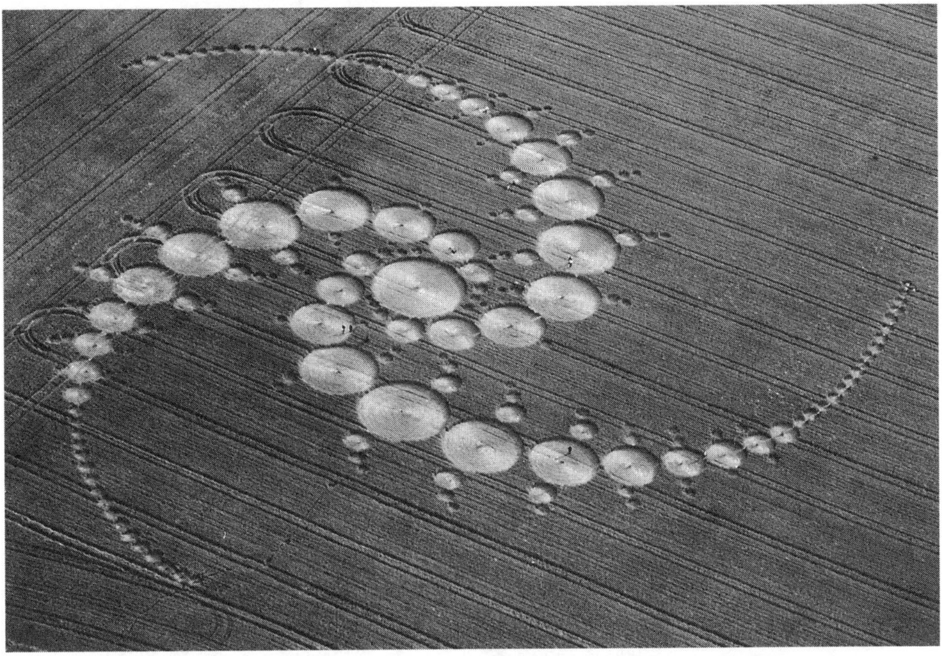

Triple Julia Set spiraled a thousand feet in diameter at Windmill Hill near Avebury. Discovered July 29, 1996, it had 194 circles. Photograph © 1996 by Steve Alexander.

I had asked in my book *Glimpses of Other Realities, Volume I: Facts & Eyewitnesses,* Could something out there be trying to reinforce the idea that our universe and all its energy and mass are defined by a repeating feedback loop that is mathematical in evolution and powered by consciousness?

Could the crop formations be a warning, a mysterious language in mathematical code from an advanced intelligence that can see the future from a timeless place, understands the repetitive patterns of our space-time, and knows mankind is at a dangerous, self-destructive moment on the evolutionary spiral? The Dutch teenager, Robbert, who has seen crops go down in the presence of mysterious lights, has the strong impression that the formations have something to do with "problems in the earth's environment." Could the intelligence(s) behind the symbols be trying to get our attention in the cereal crops of the world to warn that food and life are not guaranteed if we continue the industrial and military development that destroys so much of our planet's animal and plant life? Could human indifference to the life force that's being destroyed provoke a higher power to respond and intervene?

Another extraordinary fractal emerged in wheat at West Overton, Wiltshire, on June 24, 1999. Depending upon how you look at the pattern, you can see either ten 3-D "boxes" connected by dots, or ten flat, six-sided hexagons connected together in patterns of equilateral triangles by lines of dots also grouped in repeating triangular clusters. In mathematics and Chaos Theory, this repeating triangular pattern is related to a Sierpinski Gasket. Sierpinski was a mathematician who imagined shapes with the same bizarre qualities as the Koch Curve Snowflake.

The *Encyclopedia Britannica* explains that a classic Sierpinski Gasket starts with an equilateral triangle. Then it is divided three by three into nine equal triangles and the central one is removed. Then the operation is repeated on the eight remaining squares, putting a triangular hole in the center of each. As the process of forming this pattern continues indefinitely, the length of the curve approaches infinity around the space it encloses, similar to the Koch Snowflake.

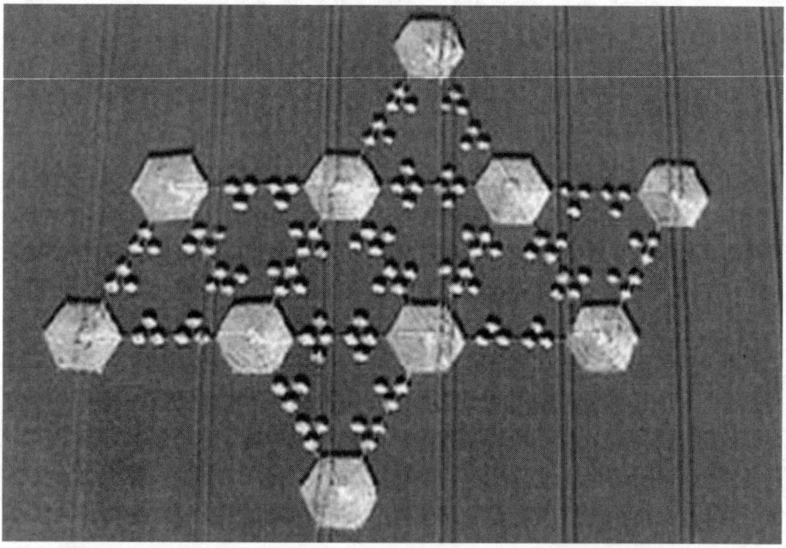

West Overton, Wiltshire, "Sierpinski Gasket" or "flat octahedron" formation in wheat, 370 feet long, discovered June 24, 1999. Aerial videograph © 1999 by Peter R. Sorensen.

A famous 3-D structure similar to this is the Eiffel Tower. Its beams, trusses and girders branch in a lattice of finer and finer detail without removing structural strength. Several people traced over photographs of this West Overton formation, cut the pattern out and folded it up into a three-dimensional octahedron.

As Frances flipped through other photographs in her binder, the 1999 formations had more three-dimensionality than past crop patterns.

"This one was close to our house." The image she was pointing to was a three-dimensional cube with very straight sides that seemed to come upward from inside a circle.

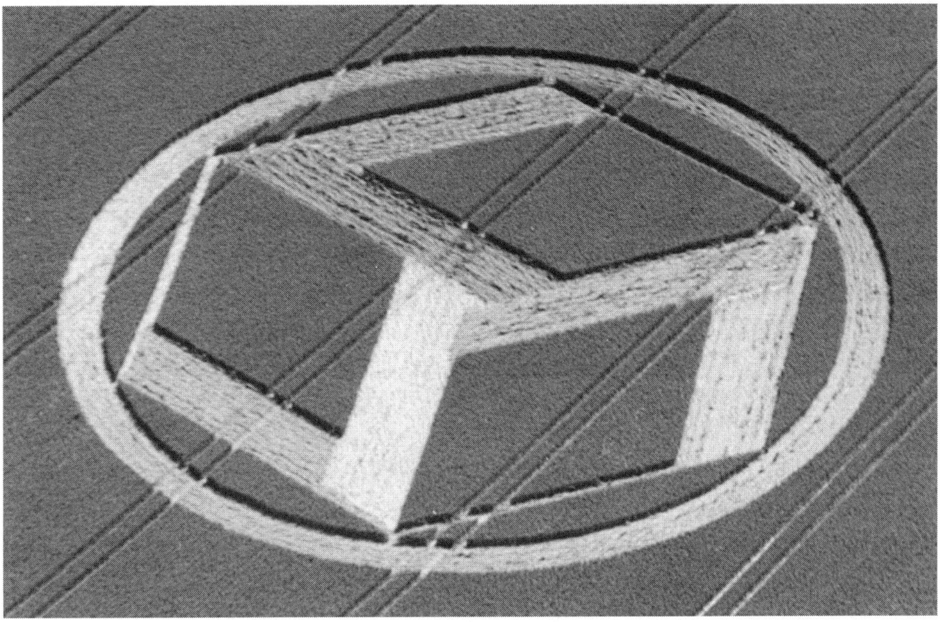

Allington, Wiltshire formation, discovered June 24, 1999, in wheat, 60.5 meters (198 feet) in diameter. Aerial photograph © 1999 by Lucy Pringle.

Frances touched her nose and forehead. "In this one, it was extremely heady. As soon as you got into it, *zap,* it got to you. As soon as you leave, the feeling goes away. There have been a lot of people who said they felt quite disoriented in that one. But it's very nice, nevertheless."

Charles agreed. "Like Frances, I felt very heady, too, almost nauseous. We visited that one three times and each time the feelings and sensations were the same. Usually formations tend to dissipate their energy and after a while, they feel like a dead, flat patch of crop. But if you catch them early, depending on your varying level of perception at the time, you can feel them. And the Allington Cube was very cleanly cut and one of the most potent crop circles I have ever visited."

Frances added, "Allington was the first cube. Then the circle makers went on to the Barge to make another cube inside a circle with a sphere inside the cube. When they first started appearing in the 1980s — Charlie moved here about a year ago, but I've lived here most of my life — they first started appearing as single circles, like a simple language. And they just progressed in complexity."

Andy Thomas, editor of the *SC Journal,* sent me information from a book, *Essential Reiki: A Complete Guide to an Ancient Healing Art,* which contains a graphic that looks like the Allington Cube, but the surface pattern does not match up. Is the crop formation a mirror opposite? In fact, if you hold this page up to a mirror, the patterns then match. Are we to interpret the formation as an *anti*-Antahkarana?

Antahkarana, a Tibetan symbol that in Sanskrit means "inner organ, inner instrument." Graphic reprinted with permission from Essential Reiki *by Diane Stein © 1995. Published by Crossing Press, Santa Cruz, California.*

Diane Stein wrote, "The Antahkarana... is a meditation and healing symbol from Tibet, mentioned by Alice Bailey and other authors as having been used in ritual for thousands of years.... It is said to be a symbol that cannot be used in negative ways; its energy has been proven by many healers for many years beyond its ancient history."

Honeystreet, Wiltshire, formation discovered July 16, 1999 in wheat near the Barge Inn, Alton Barnes, 64 meters (210 feet) in diameter. Aerial photograph © 1999 by Lucy Pringle.

Charles added, "Placing a 3-D shape on a 2-D plane seems to be suggesting 'Go deeper.' It's fairly obvious that things are getting fairly deep, fast, and we're being edged into or being asked to look into things at a much deeper level, a multidimensional level, rather than our limited way of looking at things."

I asked Charles about the Hebrew symbols that have appeared in the past few years. On May 31, 1999, at Barbury Castle north of Avebury, a Hebrew menorah appeared in barley. The base was three-pronged, as originally depicted four thousand years ago after Jehova told Moses in Exodus, Chapter 25, Verses 31–40, "Make a lampstand of pure, beaten gold. The entire lampstand and its decorations shall be one piece.…It will have three branches going out from each side of the center shaft."

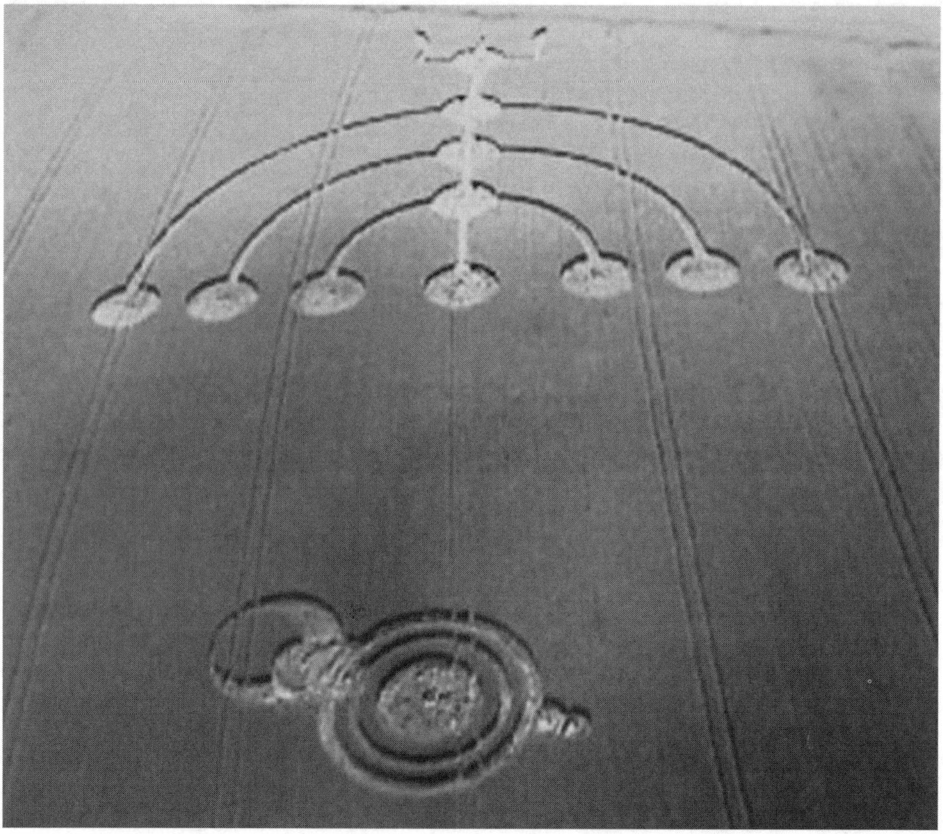

Barbury Castle north of Avebury, Wiltshire. Tripod-based menorah near horned concentric rings, discovered May 31, 1999, in barley, 300 feet long. Two years earlier, on May 3, 1997, a Kabbalah Tree of Life appeared in this same field in yellow canola. Videograph © 1999 by Peter R. Sorensen.

Hebrew scholars interpret the "one piece" of pure gold to represent the "unchanging and unified Divine World of Emanation." But it was Zechariah in the Bible (4:2–6) who asked, "What do these things mean, my lord?" referring to the gold menorah with seven lamps presented by an angel. The answer was, "Not by might nor by power, but by My spirit."

In the earliest days of its existence, the menorah's base *was* depicted as a tripod. A good example is a menorah carved into a column on the capital of the Capernaum synagogue around the third century A.D.

Menorah with tripod base carved on column in the ancient capital of the Capernaum synagogue circa third century A.D. *Photograph from Israel Ministry of Education and Culture.*

But as time evolved, most menorahs have been made and depicted with a round or octagonal base. An attorney friend in the Jewish faith, Michael Pill, who saw one of the 1999 aerial photographs of the May 31, 1999, barley formation at Barbury Castle called me to discuss the rarity of the pattern.

He said, "The depth and breadth of knowledge you have shown me in the crop formations would be very difficult to find in one individual, or even a small group. For example, the 'Tree of Life' and 'Menorah' required someone with fairly sophisticated knowledge of those subjects, especially the latter, with its historically accurate tripod base.

"I suspect that crop circles also present symbols from other spiritual/religious traditions that may not be as readily recognizable as images with roots in Judaism or Christianity. Further, creation at day or night of complex geometric designs would demand construction lines and surveying equipment. And yet there is no evidence of either in most of the formations."

The 1999 menorah with its tripod base was in the same field where the Kabbalah Tree of Life had been two years before on May 3, 1997, near Barbury Castle. The 150-foot-long Kabbalah Tree was created in yellow flowering oilseed rape (canola), a feat many think is not humanly possible without leaving construction damage in the delicate plants. When Peter Sorensen later diagrammed the Tree of Life from his video frames, he pointed out that the measurements were not proportioned the same as in most Kabbalah drawings. Why this difference, know one knows.

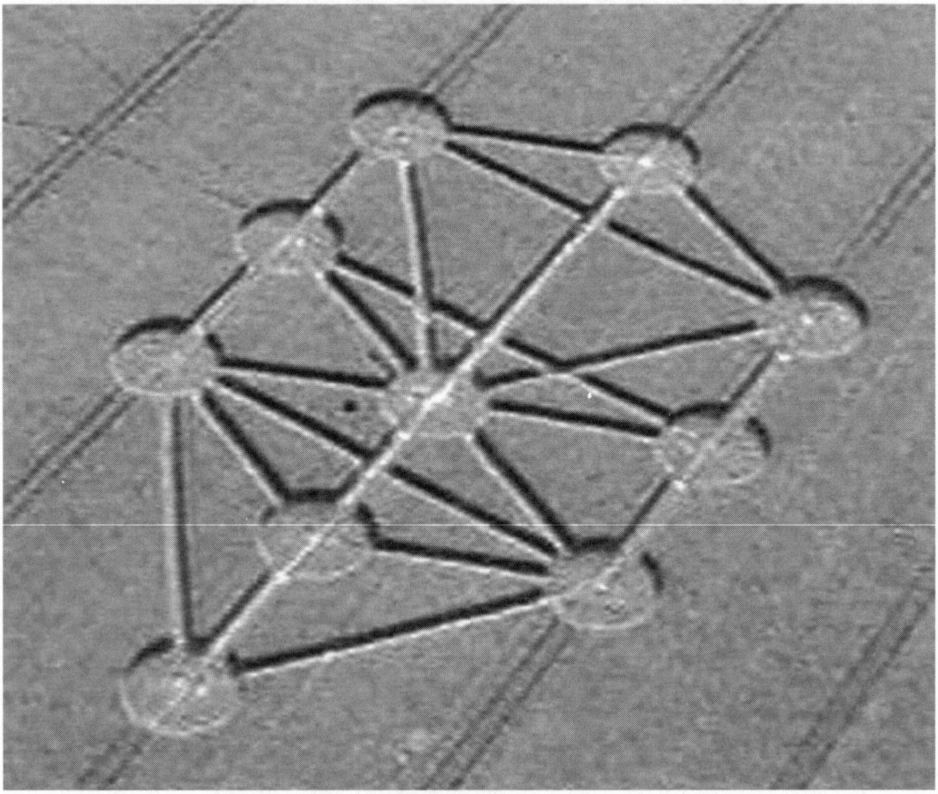

The Kabbalistic Tree of Life discovered on May 3, 1997, near Barbury Castle in same field that menorah and horned concentric rings would be found on May 31, 1999. Aerial videograph © 1997 by Peter R. Sorensen.

The Kabbalistic Tree of Life is described in *The Secret Language of Symbols* (1993) by David Fontana this way: "In order to account for the many manifestations of God, the Kabbalah contains the concept of the Four Worlds or cosmic cycles, each of which has its own Tree of Life. The Four Worlds — Manifestation, Creation, Formation and Action —

can be seen as the different aspects of God through which the universe was brought into being. They refer also to the hierarchy of the Worlds. Each world possesses all the characteristics of the one above, and so is more complex and subject to more laws....the ten sefirah are linked together by twenty-two pathways — the number of elemental letters in the Hebrew alphabet.

"Properly used and understood, the sefiroth (10 spheres) is a blueprint from which all phenomena — from cosmic forces to human relationships, from the ascent of the soul toward God to the fate of world economies — can be explained and ultimately controlled."

Some Trees of Life are also drawn with a Great Serpent entwined in the pathways and circles. The Serpent is linked to Kundalini energy that can be used to give trained individuals direct access to higher worlds.

In Manly P. Hall's 1928 book, *The Secret Teachings of All Ages,* he says that according to the Qabbalists, the life of the Supreme Creator permeates all substance, all space and all time. There is an Ain Soph, the unconditioned state of all things, the "Most Ancient of the Ancients," which no mortal eye has ever seen. The Ain Soph is symbolized by a closed eye or circle, itself emblematic of eternity. From the Ain Soph emanate "forty rates of vibration." These forty vibrations are represented as circles and arranged as four "trees," each consisting of ten circles. One could also speculate that these forty vibrations are forty dimensional existences from spirit to matter.

"These trees disclose the organization of the hierarchies controlling the destinies of all creations," Manly Hall wrote. "The trees are the same in each of the four worlds, but the powers vested in the globes express themselves differently through the substances of each world, resulting in endless differentiation." One of the four trees is named Assiah and is described as having ten hierarchies of good spirits and ten evil ones called "shells."

The continual vibration and motion of the Ain Soph "towards the center of Itself" results in the establishment of the dot in the circle called God, I AM or Eheieh. That dot and circle are the crown of each Kabbalah Tree of Life, an emblem of philosophic and soul growth. There have been variations on circles surrounded by rings of various numbers since the crop circle phenomenon began.

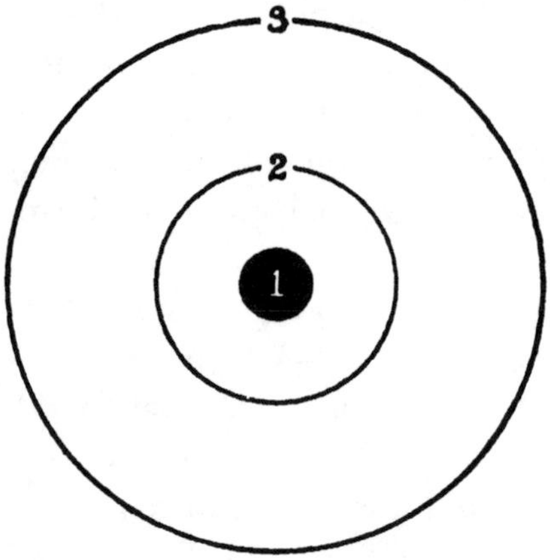

According to Kabbalists, as described by Manly Hall, "The life of the Supreme Creator permeates all substance, all space and all time. For diagramatic purposes, the Supreme All-Inclusive Life is limited by Circle 3 which may be called 'the boundary line of Divine existence.' The Divine Life permeating the area bounded by Circle 3 is focused at Point 1, 'the First Crown.' The creative forces pouring through Point 1 come into manifestation as the objective universe in the intermediate space, Circle 2.

"Out of the crown radiated nine great globes which arranged themselves in the form of a tree. All ten combined are referred to as the Sefirah," perhaps because their glowing auras have been compared to sapphires. "The ten sefirah are linked together by twenty-two pathways in each Tree of Life (which is) the number of elemental letters in the Hebrew alphabet. Each letter produces a number, an idea and a form so that mathematics are applicable to forms and ideas."

Manly Hall called this twelve-pointed diagram "Hebrew Letters according to the Sepher Yetzirah," the Hebrew Book of Creation, the oldest known Hebrew text on white magic and cosmology. It contends that the cosmos derived from the 22 letters of the Hebrew alphabet and from the 10 divine numbers (sefirot). Taken together, they were said to comprise the "32 paths of secret wisdom" by which God created the universe.

Hall described the diagram this way: "In the central triangle are the three Mother Letters from which come forth the seven Double Letters — the planets and the heavens. Surrounding the black star are the signs of the zodiac symbolized by the twelve Simple Letters. In the midst of this star is the Invisible Throne of the Most Ancient of the Ancients — the Supreme Definitionless Creator, Ain Soph."

Charles Mallett pointed out that the field next to where the menorah and Kabbalah Tree of Life appeared was where the extraordinary

1991 Barbury Castle triangle emerged. That unique formation provoked astronomer and mathematician Gerald Hawkins to report in *Science News* that five new geometry theorems were embedded in the 1991 formation that could not be found in any academic text. (See Chapter 8.)

The Barbury triangle was not in the field as late as 10 PM the evening of July 16, 1991, when a *National Geographic* television crew was taping nearby. The triangle was not all that went down in crop fields that night. A fresh circle was found at Wootton Bassett, another perfect circle was added to an existing formation at Hackpen Hill, and an enormous pictogram emerged at Preshute Down.

Eyewitnesses reported unusual lights over the Barbury field. The warden who lives on the hill "heard the most colossal roar coupled with a pulsing hum at 3:30 AM on the morning of July 17, 1991, which he described as like one hundred planes going over." The next day he found sheep scattered atop a steep hill as if they had been scared and run away from their normal pasture below.

This Barbury Castle "geometry lesson" was approximately 330 feet in diameter and was created between 10 PM, July 16, and 6 AM, July 17, 1991, the same time that several people reported seeing unusual lights over the fields. Photograph © 1991 by George Wingfield.

Brian Grist, a bookstore manager in Bristol, was out that night at Beckhampton near Avebury and a bit south of Barbury Castle. With him was his colleague, Gary Hardwick, and Gary's girlfriend. All three together watched a series of six mysterious lights move and pulse for an hour. Gary also saw for two or three seconds "a black shape, narrow at the front and wide at the rear (triangle), which obscured the stars and which passed eastward at incredible speed and totally silently."

The group also said the fifth light was baffling. Gary used binoculars and saw "a column of red, green and white pulses alternating direction from bottom to top and vice versa....Suddenly we saw a very thin beam of light join it to the ground. This beam was *curved*, not straight or jagged, and lasted one to two seconds. (Howe's emphasis. See Chapter 6 for other curved beams.) Brian Grist said, "The beam conspicuously failed to illuminate any part of the surrounding sky and appeared entirely self-contained." The group thought the beam "must have touched earth somewhere between Clyffe Pypard and Wootton Bassett."

During the hour while the mysterious lights, black triangle and curved beam were active, no sound was heard. By 6 AM, the Barbury Castle triangle was discovered by the farmer.

Brian Grist, educated in history and literature, has been particularly interested in the relationship of crop circle patterns to ancient symbols. The unique Barbury Castle formation provoked him to dig deeply into old alchemy and Mystery School texts. With his permission, here is an excerpt of what he detailed in the 1992 book *Ciphers in the Crops, The Fractal and Geometric Circles of 1991*, edited by Beth Davis.

Brian Grist: "The same triangular form also features prominently in alchemy, where it represents the triune Godhead or Trinity. See Figure B below. It will be noted that in both of these illustrations, the partitioned triangle is contained within another. How, then, do they comply with the Barbury figure, where only the one triangle was found? Close study of the figure demonstrates that the centre-points of the three outlying circular features form the points of an outer triangle which, although not etched in the corn (wheat), was an integral component of the overall design structure."

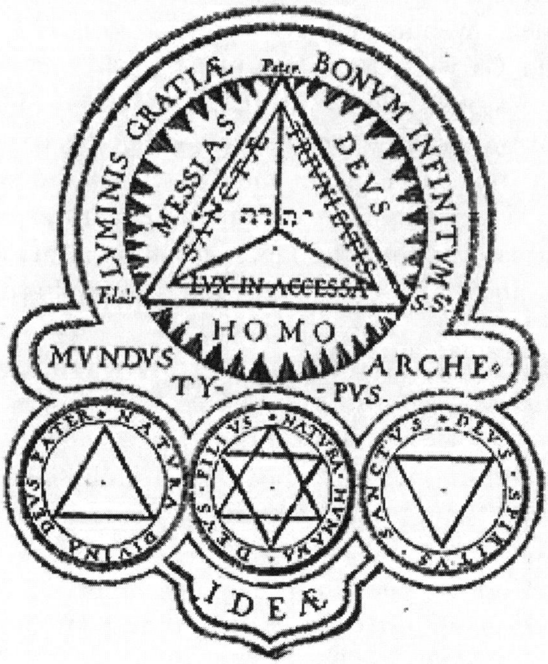

Figure B: The Triune Godhead or Trinity.

Brian Grist describes a third design (Figure C) attributed to a 15th-century monk named Basil Valentine who had the theory of *Tria Prima,* the Prime Three. "This was a fundamental tenet of alchemy which proposed salt, sulphur and mercury as the three primary elements that underlie universal matter....The most popular of Valentine's collections, *The Twelve Keys,* was first published in 1599. The twelve keys represented the twelve stages of the alchemical process. The one that interests us here is the tenth.

"The tenth key...was illustrated in diagrammatic form and was intended to designate 'the heavenly stone of the third conjunction,' merging 'Sol and Luna in the trinitarian signs of Mercurius Philosophorum, or God.' (Figure C.)

"In simple terms, this was the alchemical 'Egg' or 'Stone' in the act of formation prior to the act of completion in the form of matter. At the centre of the circle is the name of the deity, with the symbols of the sun, moon and mercury at each point of the triangle."

Figure C: Triangle of the "tenth key," attributed to monk Basil Valentine, 15th century.

The top line reads: "I am born of Hermogenes," a Gnostic philosopher noted for the assertion that "God had created all things from co-existent and unregenerated matter."

The line at right reads: "Hyperion elected me," the solar rebirth represented by the Greek sun god, Hyperion.

The line at left reads: "Without Jamsuph I am compelled to perish." Jamsuph was an unknown substance or element by which the process of rebirth was sustained and without which regeneration could not occur.

Brian Grist writes, "Here, then, we find the three essential ingredients of alchemical philosophy welded into a single image: base material, reproduction, and that mysteriously named component by which the material is transmuted."

Grist describes another triangle symbol associated with Cornelius Petraeus, a 16th-century student of Hermetic philosophy from Hamburg, Germany. Here "the alchemical process of converting base metal into gold is represented in transcendent terms." Figure D shows "SEMEN as 'seed' of the 'Great Work,'" of GOD, MAN, NATURE, SOUL and the PHILOSOPHER'S STONE.

The stone is the substance thought to have the power of transmuting baser metals into gold or, spiritually serves as a catalyst for the redemption and regeneration of humans and the universe.

Figure D: Symbol intended to represent the divine operational pattern of the creative universe with the semen seed of generation at the center.

Grist poses the question: "But how does alchemy relate to the Barbury cipher?" He found another diagram he thinks is an even closer match to the Barbury Castle triangle in Figure E. The triangle also contains that Tau/Taurus-like symbol similar to one of the three signatures at the 1998 Dadford formation discussed in Chapter 1. Figure E is from a book entitled *Cabala, Speculum Artis et Naturae, in Alchymia,* by Steffan Michelspacher, first published in 1654.

Brian Grist writes: "Starting with the unearthly beast, this fearsome creature, wearing a tiara on its horned head and spewing fire,...is a composite image of the universal cow-mother/devil-dragon: the *Anima Mundi* (soul of the world). Also named *Adam Kadmon,* this is the hermaphroditic primordial being of the Qabala: the arcane substance, no less, or undifferentiated matter. Above it we find a triangle with an overlying circle in roughly the same position as the outer ring at Barbury Castle, and at each point of the triangle, three other 'satellite' globes.

"...what merits our attention is a certain likeness between the Barbury cipher and the relationships between the Qabala's Lower Sephiroth on the Tree of Life....The Lower Sephiroth or 'circles' are the activating agents through which the Divine becomes Manifest in a relationship which, according to Qabalistic thought, permeates all living things and objects without exception.

"The world of matter is energized by the grounding of a 'lightning flash' from above, and the flow of energy between the circles; creation being dependent on their operating together as one. Each globe, or Sephira, has a different complementary function: one serves as the point of energy 'transmission'; one as the reverberating 'repeater'; one as the 'absorber'; and another, at the centre, as the point of stabilization and convergence, balancing energies from the 'celestial star' above, which is only rendered visible to mankind as heavenly light (the celestial star being the point of contact between heaven and earth).

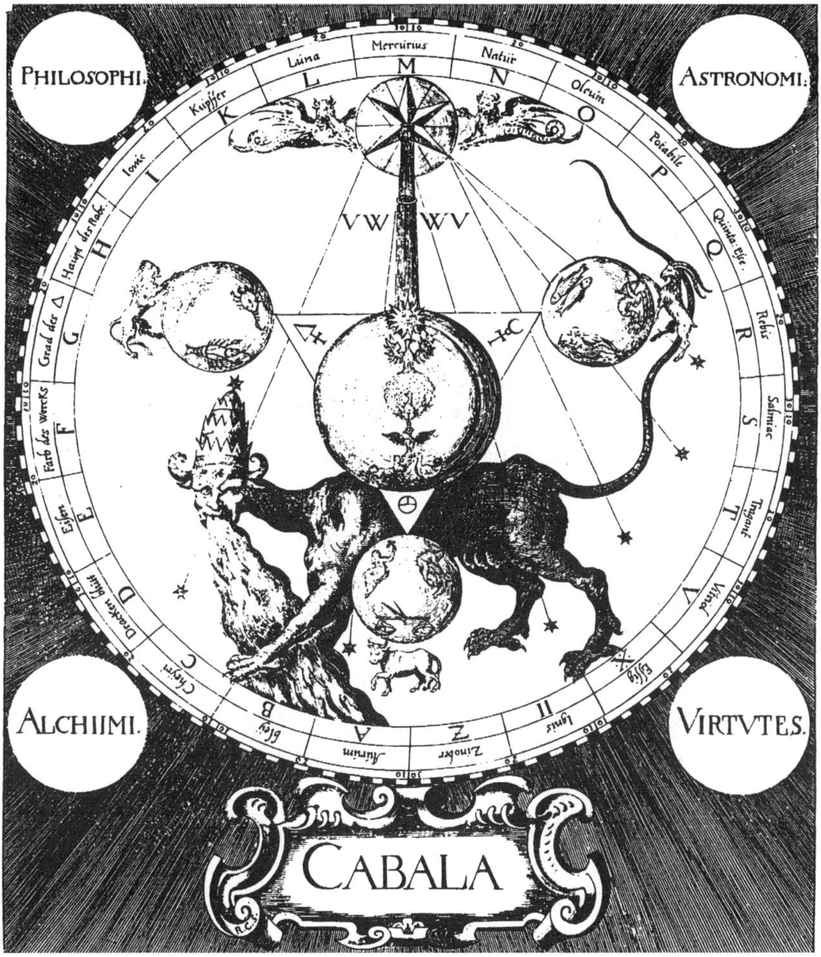

Figure E: "Exaltation" from Cabala, Speculum Artis et Naturae in Alchymia *by Steffan Michelspacher, 1654. Wheeling zodiac of good and bad signs encircling composite image of universal cow mother and evil dragon.*

Grist concluded, “Oddly enough, the crop marks at Barbury Castle appear to conform symbolically with the functions I have just outlined. The partially-pierced ring can be equated to the absorbent or ‘earth’ function; the so-called ‘sun-wheel’ with that of the repeater or radiator; the ratchet spiral with transmission.

“...one may read the Barbury cipher as a graphic representation of the ‘philosopher’s stone’ or ‘cosmic egg’ from which all life is derived, at the stage in the transmutation of matter which precedes the act of completion. The final stage is the fall into creation....

“...if the structure of the Barbury Castle formation affords us a symbolic glimpse of the interface between one level of matter and another, then we might argue that this ‘divine revelation’ has a rational equivalent in the contemporary science of Chaos Theory.”

Here we are back to fractals produced when a simple formula is repeated billions of times on a computer and creates patterns such as the Julia Sets and Mandelbrot Set called fractals, which represent an eerie mathematical order beneath a surface of seeming chaos.

Peter Sorensen thinks the Barbury Castle geometry relates to the center of the “seven thrones” diagram reprinted in Manly Hall’s book on page 46. He wrote in the 1992–93 winter *Cerealogist* published by John Michell, “The essential geometry of Barbury Castle appears at the center of the top half of the diagram. The accompanying text refers to the triangle/tetrahedron as the ‘Throne of God,’ with seven angels surrounding and guarding the throne. The three corner circles are labeled Father, Son and Spirit. The bottom half of the diagram represents the physical creation with the seven inner rings being the orbits of the then-known planets and the numerous outer rings are the paths of stars.

“My intuition is that the great glyph at Barbury Castle represents a universal truth that transcends the labels of various religions. I’ll bet that if you visit advanced beings anywhere in the Universe, you’ll find that they revere the tetrahedron as the cornerstone of creation. For our survival, it’s vital for humanity to graduate from two-dimensional thinking to three and higher dimensional thinking.”

Concerning the Tree of Life and Menorah created in the same field two years apart, Charles Mallett said, “I don’t know if there’s anything significant about all these things showing up in a fairly similar location. But the metaphysics behind the Kabbalistic stuff boils down to the essence of all religion and philosophies that these Hebrew shapes are talking about — the growth of the soul.”

Ancient Kabbalah wisdom perceived soul growth that evolves in a spiral process throughout other frequencies and dimensions back to the original source of all there is. Another Kabbalistic symbol appeared in 1999 on June 19 at Marksbury, near Bristol in North Somerset. Three overlapping circles with a triangular center were discovered in wheat.

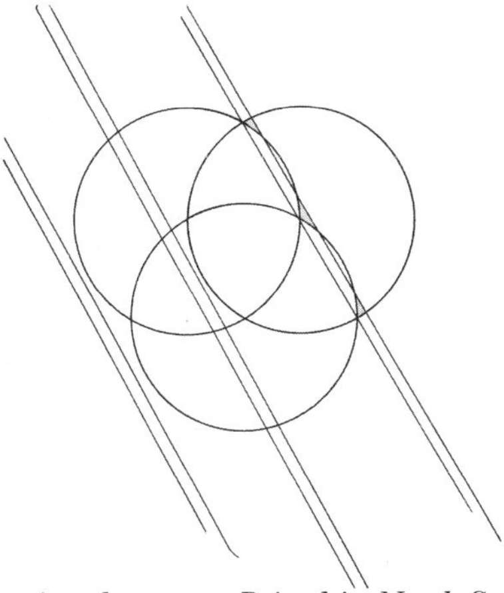

Marksbury formation in wheat near Bristol in North Somerset, England, discovered June 19, 1999. Compare to Trinity graphic below. Diagram @ by author.

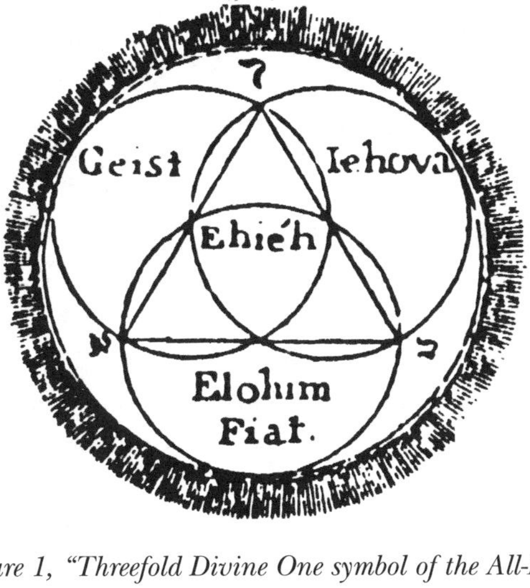

Table VII, Figure 1, "Threefold Divine One symbol of the All-Perfect, the Soul of creatures," from The Secret Teachings of All Ages, *© 1928 by Manly Palmer Hall.*

Another crop formation that seemed to have spiritual implications for most people who entered it was a six-pointed star, or hexagram, upon a hexagon inside a circle. Upon the six-pointed star was embedded a pattern of two rings and more than thirty circles of varying size arranged like spokes in the sixfold star pattern. The formation was set down right next to a Neolithic barrow known as "Devil's Den."

The hexagram of two overlapping triangles known as the "Star of David" has long been associated with the Jewish faith. However, the symbol's source is probably Arabic. Hebrew scholars such as Gershom Scholem, who intensely studied the Kabbalah early in the 20th century, concluded that "the hexagram is not a Jewish symbol. It goes back to pre-Islamic Jewish magic" where it "had one and only one purpose in its career as magic: to serve as protection against demons."

Fyfield Down formation in wheat discovered July 19, 1999, 61 meters (200 feet) in diameter. Devil's Den ancient stone dolmen is at upper right. Aerial photograph © 1999 by Steve Alexander..

Robbert in Hoeven, Holland, kept talking about deceit. Like the sigils of old, could the crop formations be part of a spiritual struggle? Could negative "shells" mentioned in the Kabbalah be part of the drama? Who is deceiving whom and to what end? Perhaps intelligences high and low are touching earth life and minds from one or more other dimensions.

If "deceit" and "antidote" are involved, as Robbert has said, perhaps symbols are being used to summon, or counteract, forces, just as black and white magicians have used symbols inside circles for centuries.

The symbol is even implicit in the ground plan of Stonehenge. Ancient science and religion researcher and author John Michell wrote about the squared circle around a hexagram in Stonehenge's 6,000-year-old ground plan, shown below, in his 1990 book, *New Light on the Mystery of Glastonbury.*

Michell wrote, "The square, drawn with its perimeter equal in length to the mean circumference of the outer sarsen circle, contains the circle of 'bluestones.' Within this (implied), a six-pointed star contains a circle of the same width as the inner bluestone formation."

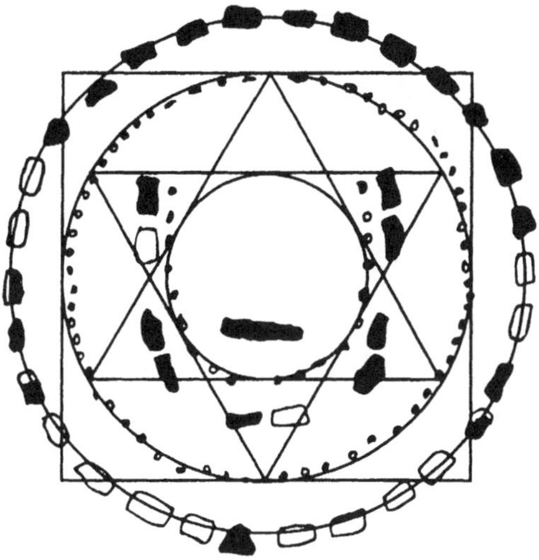

Stonehenge ground plan of large sarsen stones drawn with superimposed hexagram. From New Light on the Mystery of Glastonbury, *© 1990 by John Michell.*

Leonora Leet says in her 1999 book, *The Secret Doctrine of the Kabbalah, Recovering the Key to Hebraic Sacred Science,* "It has long been thought that the hexagram was [originally from] Arabic alchemy, [but it was] the West [which] most popularized this symbol of equilibrium between the opposed spiritual forces of fire (expansive force represented by the upward-pointing triangle) and water (the contractive force represented by the downward-pointing triangle). Since it was only in the early Middle Ages that the hexagram began to be featured prominently in Jewish magical texts and amulets, most modern historians have concluded that it was probably through the Arabic source that the hexagram was most directly transmitted to the medieval Spanish Kabbalists."

Could one symbol be laid down for a purpose that is challenged or neutralized by another symbol? Could we better understand the crop drama if we studied geometry and ancient magic rituals?

Or could crop formations serve several purposes at once? In addition to whatever force the symbols might assert, could their geographic placement provide grid points or even monitors the way seismometers indicate earthquake tremors?

I asked Charles and Frances Mallett what they thought about the Devil's Den formation.

Charles said, "I think the Devil's Den at Fyfield Down is one of the most amazing formations of 1999. Major camera anomalies in that one for myself and lots of other people. The first time we visited, I think it was the day it arrived. It was very fresh. Very amazing. It was like a maze inside there. At ground level, the pattern was nearly impossible to figure out. My camera stopped working. I walked out of the formation and up the tramline and 20 meters outside the crop circle, the camera is fine. I take my pole shots from outside. I walk back into the crop circle, and my camera won't work again. I walk up the tramline, and it would start again."

Thinking of Dr. Levengood's work, I said, "As if a residue of whatever energy made the formation is still there and can affect the electromagnetic fields."

"Yes, you can pretty much detect the energy field by dowsing. Walk into a fresh crop circle with dowsing rods and as soon as you hit the edge, they'll cross or go apart. Anyone can do it."

A few days later after talking with Charles and Frances, I spent half a day trying to find the Fyfield Devil's Den formation. It could only be seen from the air or a hillside above it. There were no roads or public access to the field. I found a farm manager and asked permission to walk across his land. He was happy to help and before I took off, I asked him, "Why is it called Devil's Den?"

He answered, "Most people stay away from it, but I heard there's an imprint of the Devil's palm in the ceiling of that rock."

He was looking at me soberly from pale blue eyes.

"Do you believe that?" I asked him.

"Don't know. Never went there to look and don't think I want to."

Huge shadows moved across the valley one after another as summer cumulus clouds were pushed along by upper winds. But on the ground as I walked, there was only a slight breeze in the hot air. I had to walk nearly a mile to reach the six-armed star next to the rock dolmen.

Fyfield Down and the Devil's Den dolmen of sarsen rocks to the right, August 5, 1999. Photograph © 1999 by author.

I could stand under the rock roof and was surprised to find two spirals carved in the old stone. Both spirals had a thin layer of moss that had turned them slightly green. Such spiral labyrinths have been found carved on ancient rocks in England, Ireland, the Adriatic Sea region, Hopi Indian lands, Greece and the island of Crete dating as far back as four thousand years before Christ.

Some historians think labyrinths and spirals marked sites of initiation to new knowledge. Their symbolic meaning has been associated with the cyclic renewal of life, the great round of death and rebirth, the journey of the soul in and out of past, present and future forms.

In my 1998 book, *Glimpses of Other Realities, Volume II: High Strangeness,* I described an experience with light I had as a teenager in Idaho mountains that resonated with the spiral labyrinths.

I wrote, "Since then, I have been convinced that a common source of energy pervades all there is and can be an ally to living life. That energy, I think, moves in cycles like the spiral labyrinths found carved in rock all over this planet. Part of the old symbol's meaning is that the machinery of the universe involves the evolution of souls. The spirals are not static and not two-dimensional. The symbol is a slice. A larger reality can be imagined as a spiral upward and downward through innumerable other frequencies or dimensions.

"Ancient wisdom understood that the moment of death was simply a transition into another frequency on the spiral. So all lines in all directions are simultaneous and filled with life forms ebbing and flowing, supported by a singular force, an invisible matrix of energy from which everything emerges and to which everything returns."

Hocky Valley Labyrinth, Tintagel, Cornwall, England. This labyrinth design is one of a pair carved on the rock face beside the river. This symbol has also been found carved on ancient rocks in Ireland, the Adriatic Sea region, Hopi Indian lands, Greece and the island of Crete dating as far back as four thousand years before Christ. The labyrinths have become a universal symbol of the cyclic renewal of life, the great round of death and rebirth, the journey of the soul.

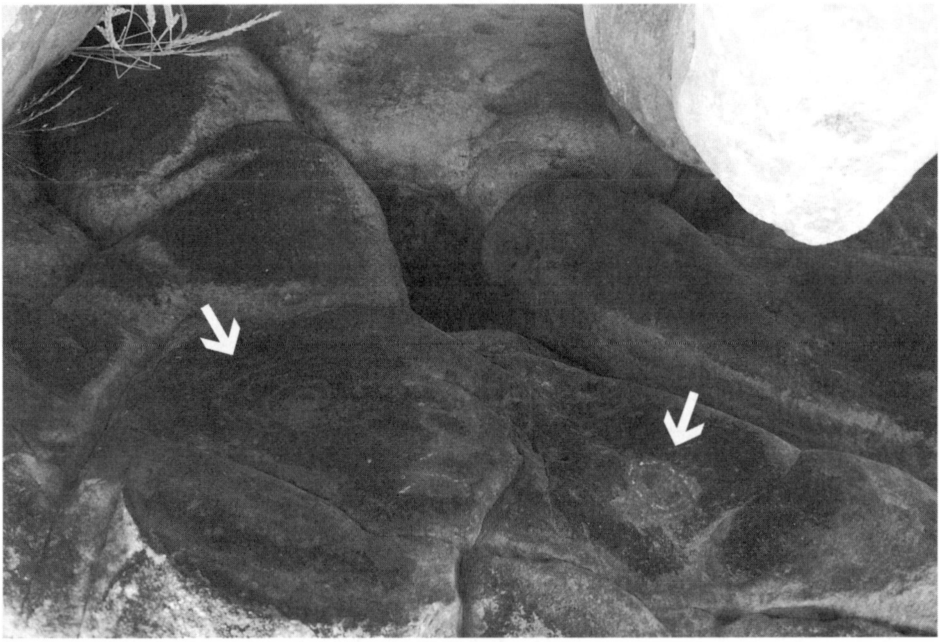

Inside the Neolithic "Devil's Den" at Fyfield Down near Marlborough, Wiltshire, looking up at alleged "Devil's Palm Print" on underside of rock roof. Arrows point to two moss-covered spirals.
Photograph © 1999 by author.

Close-up of largest spiral carved in ceiling of Devil's Den.
Photograph © 1999 by author.

The ancient Hindu tradition also described a life force called Sakti, or the "energy of the gods," which was thought to materialize in spiral form and bring with it the forces of attraction and repulsion.

In 1985, neurosurgery researcher Richard Bergland wrote in *The Fabric of Mind:* "Wise men from India predicted that mystical forces regulated the activities of the brain and body. Their view of Karma as a life-giving force that flows from life to life is in every way like the modern physicist's view of electrons (electricity) which they say is not 'matter,' but a 'force.' The Hindu commitment to reincarnation is a wager that Karma moves from one moving life to another, moving life by the same 'magic' that moves atomic electrons between orbits."

A year earlier, on June 21, 1998, in a barley field near the Devil's Den, a formation appeared with four overlapping circles centered around a "slit" referred to as a "Cat's Eye." The sections were defined by alternating layers of crop laid in different directions or standing.

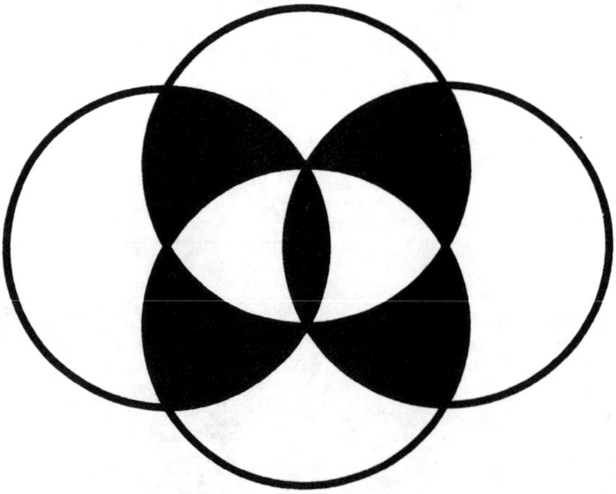

Devil's Den formation in barley, about 150 feet in diameter, June 21, 1998. Diagram © 1998 by Peter R. Sorensen.

The next day, on June 22, 1998, another similar formation was found in an Alton Barnes field, but the laid and standing crop created a pattern of "colors" opposite to the one near Devil's Den. This one had the addition of four tiny grapeshot where each circle overlapped, the iris was smaller and many plants were oddly bent halfway up the stems.

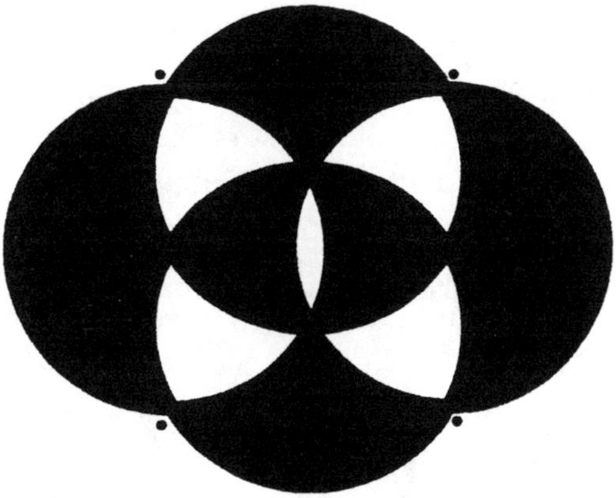

Formation in barley north of Alton Barnes June 22, 1998. Diagram © 1998 by Peter R. Sorensen.

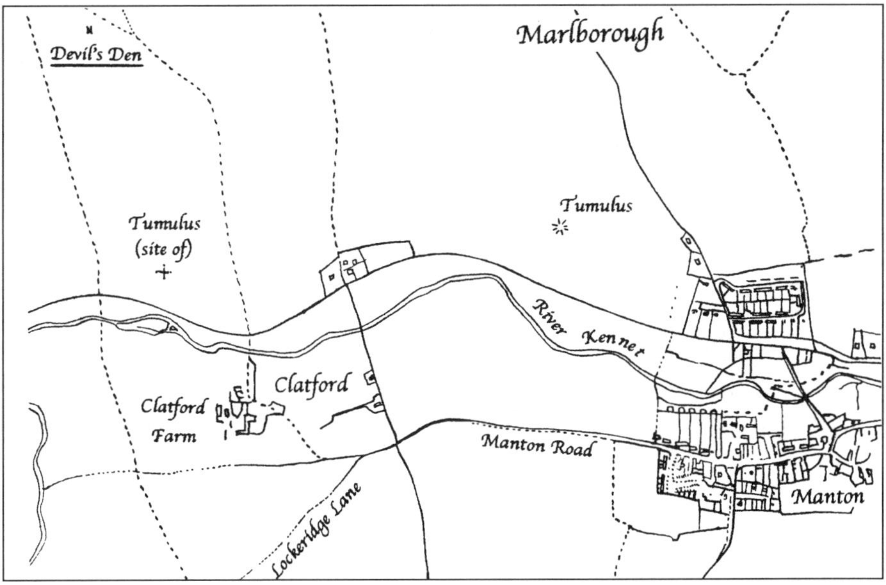

Location of Devil's Den Neolithic rock dolmen (upper left) west of Marlborough surrounded by farms. Map © 2000 author.

Back at the hill above Roundway, the last photograph that Charles and Frances showed me in their album was a rectangle inside a circle, discovered at Windmill Hill near Avebury on July 16, 1999. The couple said that the standing crop that formed the center pattern reminded them of Knight Templar designs used in churches and buildings.

Charles said, "We had friends up on Windmill Hill that night with night vision equipment. They were up until 2 AM, but didn't see a thing."

I was studying all the circles in the photograph. There were 288 of them, breaking the record for the most circles ever found in one formation. The formation reminded me of an integrated circuit.

Charles stared at the photo. "Right, could be that, too."

Another pattern similar to Windmill Hill, a kind of mirror image like the two cat's eyes from 1998, was yet to occur at the West Kennett Long Barrow. (See Chapter 4.)

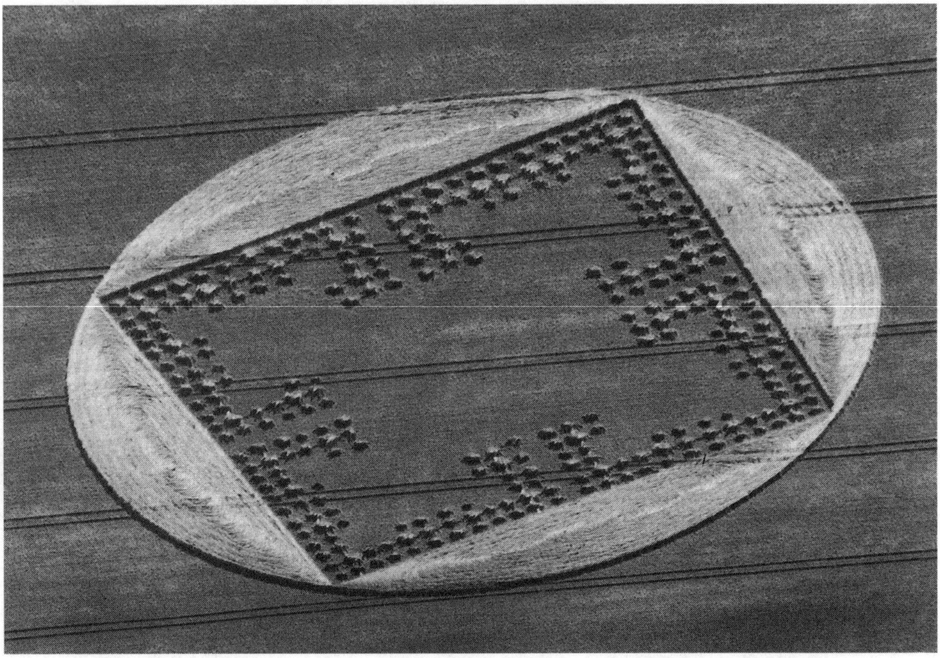

Windmill Hill, Wiltshire, near Avebury and Silbury Hill, discovered July 16, 1999, in wheat, 100 meters (328 feet) in diameter, 288 circles. Aerial videograph © 1999 by Peter R. Sorensen

At 5 PM the sun was still high in the English summer sky, but Frances and Charles needed to get home to their children, and I wanted to get into the Roundway formation. I stood at the top of the hill staring down at the amazing 360-foot design. The seven large outer circles and seven small inner circles underscored a 1999 theme of sevens and sixes. More than a snowflake, there was a regal quality to the design, "almost like a crown," Frances said.

I started to move down toward it and suddenly remembered I was supposed to meet Nick Nicholson, editor of *The Circular Review*, to talk about his own recent experiences with mysterious lights. Nick had traveled from his home base in Calverton, Nottingham. I would come back to Roundway later. I headed for the one place in Wiltshire you can always catch up on the latest photographs, new formations and news in general: the Barge Inn on the Kennet and Avon Canal at Alton Barnes.

3
The Barge

The Barge Inn in Alton Barnes, Wiltshire, England, since 1810.
Photograph © 1999 by author.

Nick Nicholson was waiting for me at the bar with his friend, Carl Nevin, another witness of the mysterious lights they had videotaped and wanted to show me. Also with them was Wolfgang Schindler, one of the first researchers to sequence crop formations into a pattern. Wolfgang had discovered repeating pentagonal (five-pointed star) geometries that convinced him there is a keen intelligence behind the crop formation phenomenon.

We decided to go outside along the beautiful canal where we could talk on audiotape while we enjoyed the extraordinarily dry, hot last day of July.

Nick Nicholson has been coming to Wiltshire from his home in Calverton, Nottingham, England, since 1990. There he drives a truck for a small company, but his passion is to understand the who, what and why behind the crop formations. He began editing *The Circular Review* in 1995 and in the summer of 1999 had his own glimpse of mysterious lights.

Barges with sleeping quarters carry residents and tourists up and down the Kennet and Avon Canal between Reading and Bristol. The Alton Barnes White Horse carved in 1812 is on the hill in the background. Photograph © 1999 by author.

Carl Nevin, Wolfgang Schindler and Nick Nicholson, editor of The Circular Review, *outside the Barge Inn on Saturday, July 31, 1999. Photograph © 1999 by author.*

One of the witnesses with him was Carl Nevin, a factory worker in domestic and military electronics from Parkside, Stafford, England. Nick began explaining what happened.

"It was only four nights ago on July 28. We arranged for a group to go on a sky-watch up to Hackpen Hill, which is the highest point in the Avebury to Swindon area. We overlooked the Hackpen formation that sat in the field down below. It's like a spinning set of blades marked out in the crop. The farmer also saw strange lights over the field the night it came."

Andy Thomas, editor of *SC (Sussex Circular): The Bimonthly Journal of Crop Circles and Beyond,* wrote that on the night Hackpen formed, "researchers from the cropcircleconnector.com web site were actually carrying out a cropwatch from the car park on the hill.

"Shortly after midnight, small lights and what looked like a searchlight beam scouring the fields below were witnessed by the group. And the next morning the formation lay below."

Nick added, "Now, Andreas Mueller will tell you that he thinks it could be depicting the saros cycle, which has to do with the solar eclipses." (See pages 141–143.)

The first English formations in April 1999 began with an eclipse motif in blooming yellow canola plants, also known as oilseed rape, which led to anticipation about the August 11, 1999, total solar eclipse.

Great Britain had not had a total solar eclipse cross its land since 1927. The next will occur in 2090. I told Nick that I planned to talk with Andreas Mueller about his survey work and his discovery of a saros solar eclipse graphic that he compared to the Hackpen Hill arcs.

Hackpen Hill formation in wheat, 120 meters (394 feet) in diameter discovered July 4, 1999.
Aerial photograph © 1999 by Lucy Pringle.

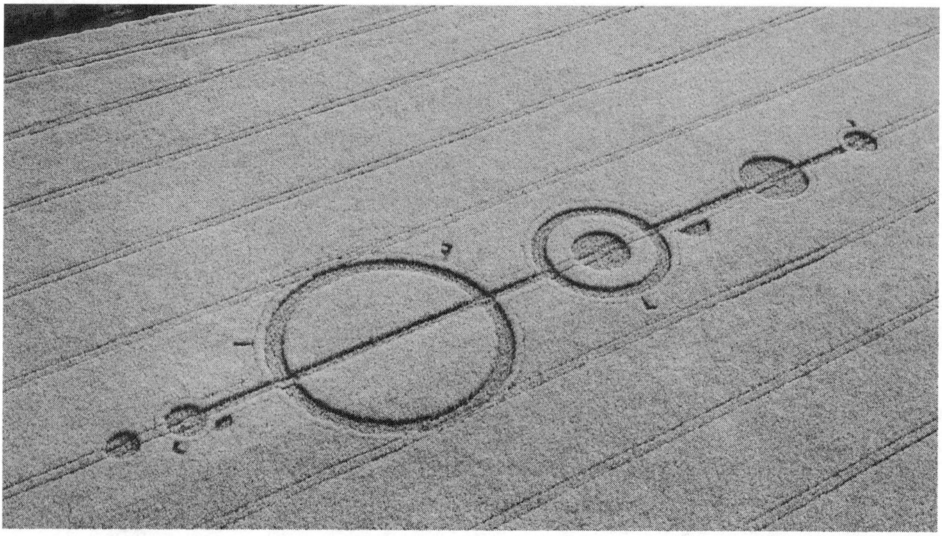

Over Wallop, Hampshire, discovered April 3, 1999, in blooming yellow oilseed rape, 300 feet long. Aerial photograph © 1999 by Steve Alexander.

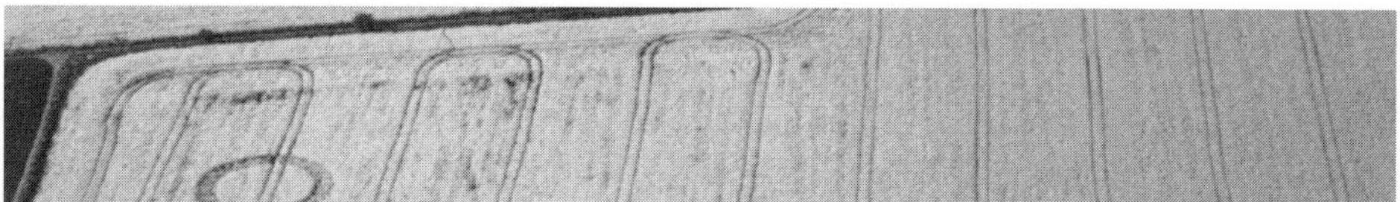

Milk Hill, near Alton Barnes, Wiltshire. Five "eclipse" circles on left discovered April 4, 1999, 400 feet long; pattern on right arrived May 2, 1999, 250 feet long, in which Charles Mallett encountered the sphere of light. Aerial photograph © 1999 by Steve Alexander.

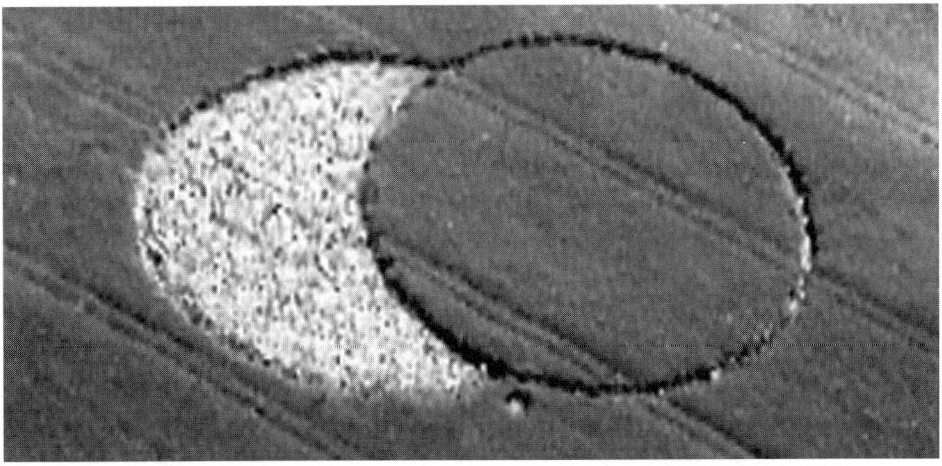

"Solar eclipse" at Bishops Cannings, Wiltshire, discovered April 16, 1999, 100 feet long. Canola blooms strangely delayed in bright yellow crescent and tiny circle at overlap. The rest of the field had already bloomed. Aerial videograph © 1999 by Peter R. Sorensen.

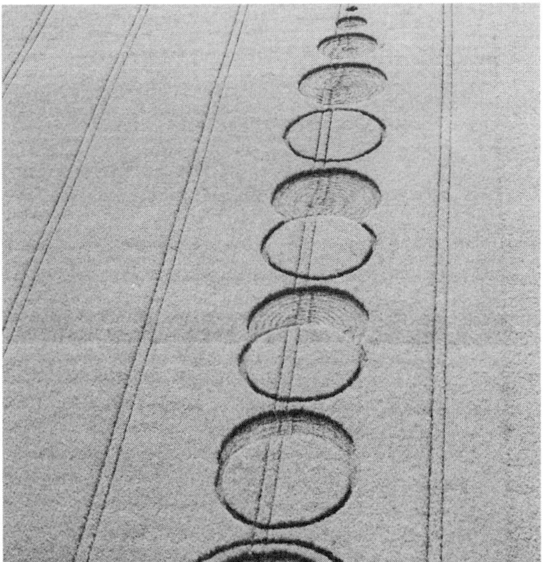

Middle Wallop, Hampshire, discovered May 2, 1999, in blooming canola, 350 feet long. Aerial photograph © 1999 by Steve Alexander.

The fact that only the crescent and tiny dot were in the yellow flower stage, compared with the rest of the field which had already bloomed several days before, implies a retarded growth rate specifically in the crescent and tiny circle plants. If so, energies affecting those plants would have interacted *weeks before* the formation itself appeared, as Dr. Levengood also hypothesized about the Avebury and Whitefish, Montana, cases discussed in Chapter 8.

Perhaps a blueprint of the formation-to-be had been superimposed by applied energies on the plants much earlier in the season in order to retard growth in the specific canola that would form the crescent and tiny circle so they alone would be yellow flowers next to the older, green canola circle. But whose blueprint? And what energy system?

Whatever energies are interacting with the cereal crops and grasses of the world, lights, translucent spheres and flickering brightnesses seem to be associated with them. I asked Nick about the orange lights he, Carl Nevin, Mark Hayward and others saw and photographed on Hackpen Hill the night of July 28, 1999 — the same night that Charles Mallett photographed an orange light near Avebury. (See Chapter 2.)

"Right. Well, from up there, it's an excellent 360-degree view over all of the countryside. Behind us was a tall radio mast that sits on what is known as the Ridge Road about three miles away. And that radio tower is marked by a red light that does not flash. It's a constant light. Further to the right of that, there is the top of another radio tower which on subsequent investigation proved to be a mast that sits on the other side of the M-4 motorway toward Swindon. That's got multiple red lights of which we could see three at the top of the mast.

"It was almost a full moon that evening, so the stars were a little bit obscured by moonlight. But prior to our arriving, some other people had watched an orange light towards the eastern horizon that had been performing all over the place, wandering around. But unfortunately, they didn't have any video equipment.

"Before then, we were at the Avebury Avenue of Stones around 9 PM when orange lights showed up there and we got those on tape, too. We could see them first moving together and then slightly apart. One light was a bit brighter than the other one. Eventually, one light went out first. They just go out. They don't descend.

"The other orange light stayed there, but you could actually make out through the binoculars that the orange lights themselves were not constant illumination. They were pulsating. You couldn't see where they were attached to anything at all. But you could see the actual light itself was pulsating. They were about five degrees above the horizon, so they weren't a vehicle on a road in the area."

Carl Nevin agreed. "The orange lights were high enough to not be supported from the ground."

"And Carl, you could see the lights move, too?"

"It seemed to be just horizontal movements. They were first apart, and then they suddenly seemed to come close together. Then the one on the right seemed to get more dim."

"What geographic area did the lights seem to be over?"

Carl answered, "The only place near was the formation at Liddington Castle which arrived a few days before."

Videotape Frame 1: Orange and white lights seen and photographed from Avebury Avenue of Stones, Wiltshire, on July 28, 1999, just before 9 PM. Video frame © 1999 by Mark Haywood and Carl Nevin.

Nick added, "It was not in the direction of Avebury. We have positioning compasses with us and they read off ninety-five degrees east of true north, which made them virtually due east from where we were towards Marlborough or between Marlborough and Swindon."

Nick said, "I've interviewed other people who guess these things are still around after they pop out, but people don't generally look at the blank space with binoculars to confirm if something is still there. So, the fact that through binoculars one of the lights that seemed to disappear was still faintly there, I think, is significant."

Videotape Frame 2: Two seconds later after video frame 1, the lights have moved. Video Frame © 1999 by Mark Haywood and Carl Nevin.

Carl got his videotape camera and set it up so I could see what he photographed after Avebury at Hackpen Hill. I have considerable experience as a TV producer, documentary filmmaker and editor, and have seen hundreds of videotapes taken by people all over the world of unusual lights in the sky, sometimes even structured craft. A big problem in most of them is no tripod. People are excited and try to videotape a small light source or aerial object while holding a lightweight camera that shakes and moves with every breath. Carl videotaped without a tripod. But through the viewfinder, I could clearly see a light moving in the frames.

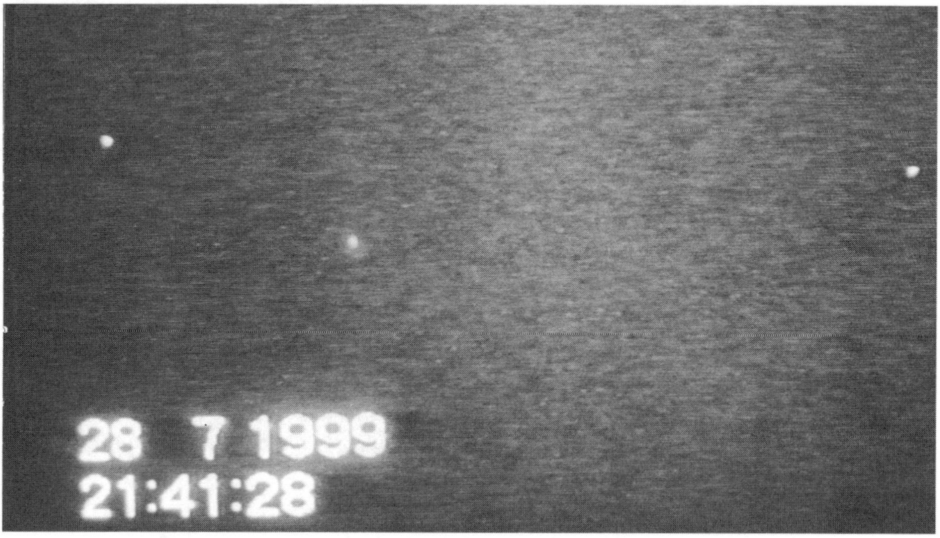

Video Frame 3: At 9:41 PM, *red and orange lights appeared a second time at Hackpen Hill. Video frame © 1999 by Mark Haywood and Carl Nevin.*

Video Frame 4: Fifty-one seconds later, Carl Nevin zoomed the lens in where the far left light has disappeared down to "a very fine point in binoculars" while the far right light moved closer to top frame center above bright red light in the center. Video frame © 1999 by Mark Haywood and Carl Nevin.

"Nick, how long were the Hackpen lights visible?"

"Well, the second time the orange lights appeared at Hackpen was a little before 10 PM. Everyone saw the second round, including us. And although to the naked eye one of them disappeared totally, it could still be seen with binoculars as a very fine point of light."

Then Carl showed me his maximum digital zoom on the orange light at seventy-two magnification. The light not only pulsed, but it seemed to be spinning.

"It could be spinning or pulsating," Carl said. "Very difficult to focus in on a long-distance object. Some of that movement is mine and some could be the actual object. I'm zooming in and out to give you the maximum *optical* resolution and then into the *digital* resolution where you can see a lot more going on."

Photographer Peter Sorensen cautions, "Digital zooms can introduce artifacts and should not be used for documentation. Digital zooms can be done later in a video production house if needed. It's better to stay on normal wide shot or normal zoom lens during an event."

"Carl, does the light do any forward and backward motion on this tape versus your zooming the camera?"

"Some of it was me," Carl answered, "but over a period of time you'll see this light go dimmer. It appears to be going back and then coming forward separate from me zooming back in again. I was panicked to get the best out of it. Here is the important part. I have remained on maximum pixilation seventy-two times digital. I've left it fixed. Now you will see this thing suddenly goes into the distance. See?"

"Yes." The light in the black frame did seem to suddenly grow smaller and dimmer as if moving away from the camera, or perhaps dimming itself from the perimeter inward to the center.

"I didn't touch the camera at all. That's the object itself reversing back into the distance, if you can call it reversing. It's gone!"

Nick Nicholson also had other photographs of mysterious lights he shared with me, and thought they would contribute to the overall pattern of reports he and others in England have received in association with crop formation activity.

"Photograph 1," he said, "was taken at Goodworth Clatford in Hampshire in 1995 by a man who was there the same time I was and later gave it to me, requesting anonymity. Nothing was seen at the time of photography by either myself or Mark Haywood who was also with me. The image only became apparent after the film was developed. The lab who carried out this operation stated that there was no fault in the film and the aberration was part of the image.

"This 1995 formation where the light was photographed was a very strange and unique event. It had at last count forty-four different randomly placed grapeshot on the western side of the field. From the ground these were not apparent. Airborne, the effect was amazing. Some were like small formations themselves."

Mysterious light image on photograph 1, but not seen at the time near crop formation with forty-four grapeshot in Goodworth Clatford, Hampshire, July 1995. Photograph provided anonymously to Nick Nicholson © 1995.

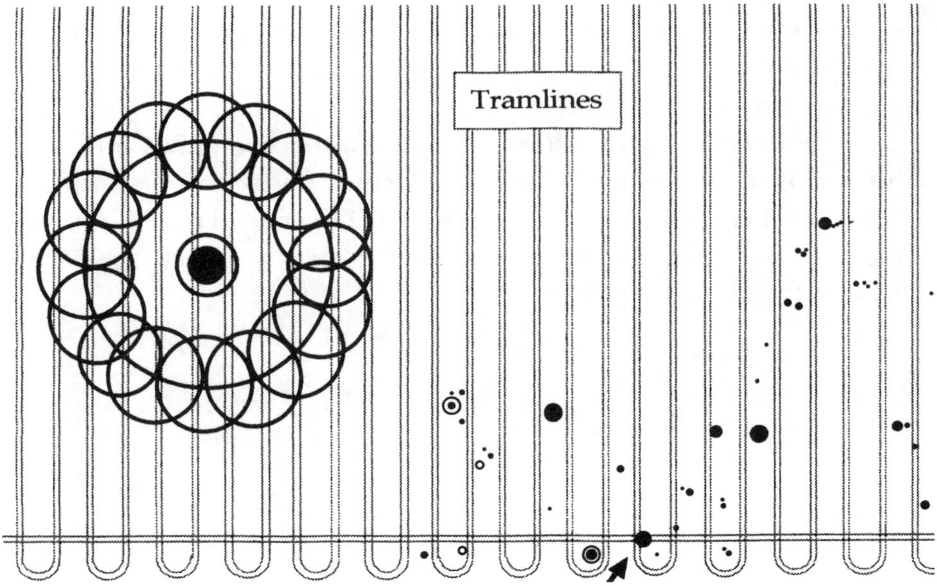

Goodworth Clatford south of Andover, mid-July 1995, complex ring formation in barley, 370 feet in diameter. Nearby were at least forty-four small grapeshot circles. Several were in the middle of tramlines with no paths or footprints leading to them. Diagram © 1995 by Peter R. Sorensen.

Nick said, "I had revisited the field for ground photographs of these grapeshot. Photograph 2 is my own that also shows a strange white, misty light in the left-hand corner that seems to spread out. The Goodworth Clatford formation I went to photograph can faintly be seen in the field beyond the mysterious light. Alongside the field was a water treatment plant, and Colin Andrews remarked that the formation seemed to reflect the action of that contraption in the series of interlinked rings."

Photograph 2 shows another mysterious misty glow in the same Goodworth Clatford formation. Photograph © July 1995 by Nick Nicholson.

In mid-May 1995, another formation appeared at Hazel Down south of Goodworth Clatford, fifty feet diameter in oilseed rape, that seemed to be a variation on the Trinity symbol.

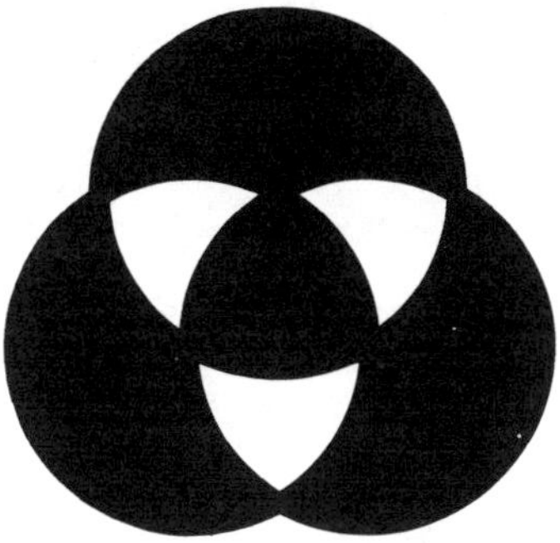

Mid-May 1995 at Hazel Down south of Goodworth Clatford about 50 feet on a side in oilseed rape. Diagram © 1995 by Peter R. Sorensen.

In mid-June 1996 near Goodworth Clatford, a barley field had a beautiful "rose" formation. Peter Sorensen thought it also resembled a cymatic pattern that is produced when liquid surfaces are vibrated.

"Rose" or "Cymatic" vibration pattern in barley, mid-June 1996.
Diagram © 1996 by Peter R. Sorensen.

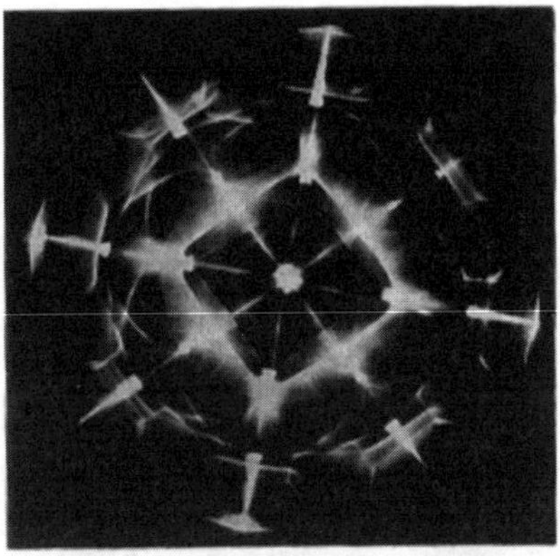

Actual photograph of pattern produced by vibrations in oil of turpentine, from Serpent in the Sky, *© 1993 by John Anthony West.*

Egyptologist and writer John Anthony West wrote in his 1993 classic *Serpent in the Sky,* "Cymatics, the study of wave forms, illustrates dramatically the relationship between frequency and form. Specific materials subjected to specific vibrations assume specific forms. A given form can *only* be summoned forth at its corresponding frequency; form is a *response* to frequency. Form is what we call 'reality,' but that reality is obviously conditional — for it is the structure of our organs of perception that is responsible for the ultimate picture. If our senses were differently attuned, reality would assume a very different aspect.

"...Vibration is an alternation between positive and negative poles. Metaphysically, it is a manifestation of the revolt of spirit against its imprisonment in matter. Cymatics gives visual expression to 'In the beginning was the Word.'"

And on May 10, 1998, a three- "petalled" formation emerged in oilseed rape that was about 140 feet in diameter.

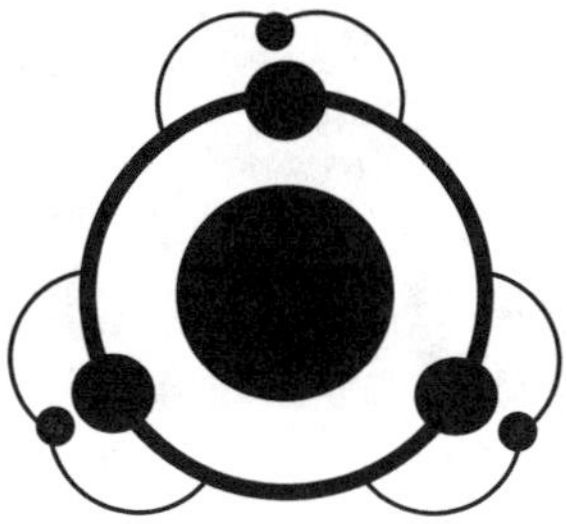

"Tripetal" formation discovered May 10, 1998, in oilseed rape near Goodworth Clatford, Hampshire, about 140 feet across. Diagram © 1998 by Peter R. Sorensen.

A decade before in 1985, Goodworth Clatford, Hampshire, was one of the first areas to have a Celtic cross pattern. Nearby, a flattened circle was surrounded by two rings, but one side looked as if the formation had collided with a wall.

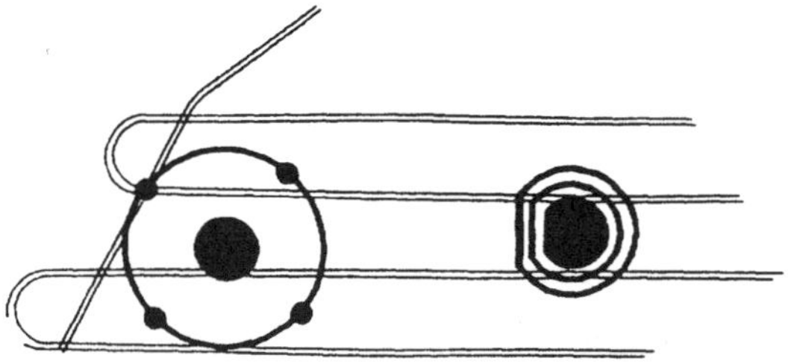

1985 Celtic Cross and blunted ringed circle near Goodworth Clatford.

In mid-August 1999, there was another formation at Goodworth Clatford. No one surveyed it, but pilot and crop researcher Busty Taylor flew over and made an aerial photograph. The 1999 Goodworth Clatford formation had a similar style to three others discovered farther southeast between East and West Meon. (See Chapter 4.)

Goodworth Clatford near Andover, Hampshire, reported mid-August. Aerial photograph © 1999 by Busty Taylor.

The July 1995 misty light at Goodworth Clatford wasn't the first Nick Nicholson had seen and photographed. In 1991, he had photographed similar odd, misty lights near a Tawsmead Copse formation in Alton Barnes that he labeled photograph 3.

"Photograph 3," Nick said, "is from the lower road running past the East Field. As far as I was concerned, the intention was to capture an image of the formation you see in the upper part of the frame. Again, no lights were apparent to me. There was also no evidence of an aberration on the videotape which I took at the same time. The lights show up only on the photographic film. The white misty lights remain a mystery to this day."

Photograph 3 is from the lower road running past the East Field at Tawsmead Copse in late afternoon and shows unidentified misty lights with a formation beyond. Photograph © 1991 by Nick Nicholson.

"Another similar misty, white light showed up in photograph 4 taken at a 1995 Roundway formation by Denni Clarke from Freedom, California. She did not see anything unusual at the time either." A photograph of this 1995 Roundway formation is in Chapter 4.

Photograph 4 shows another strange, misty light coming up from the center of the wheat at a 1995 Roundway formation. Photograph © 1995 by Denni Clarke, Freedom, California.

Two years later, in early August 1997, a pictogram of five circles and connecting paths was discovered in an oat field near Marion, New York, in which a similar misty anomaly was photographed.

On the afternoon of August 24, 1997, crop circle researcher Larry Thomas visited the pictogram of five circles and connecting paths laid down in oats. As Nick Nicholson had done in England, he took photographs without noticing anything unusual at the time. But on one of his prints was a strange, gray "smoke" seeming to come right out of the oats.

Larry told me that a woman had been walking her dog around 11:30 PM the night of August 22 when she noticed "a slight humming noise and a bright light. She looked up and saw an object hovering directly over the field where the crop formation later formed. The object was about three cars wide and had a distinct overall appearance of an acorn in shape. It had a domed roof with a distinct ridge or band running around the center. And it was about five hundred yards from her when she witnessed it."

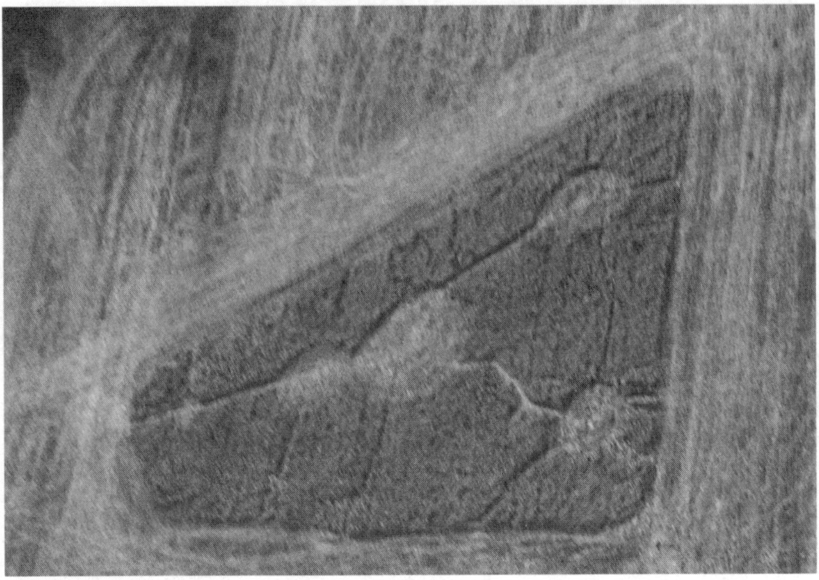

August 1997 aerial of oat field, five circles connected by pathways in Marion, New York, pictogram. Aerial photo © 1997 by Larry G. Thomas.

Strange "misty glow" was not visible when this photograph was taken on the afternoon of August 24, 1997, in the Marion, New York, oats formation. Photograph © 1997 by Larry G. Thomas.

Larry gathered soil from that 1997 formation to have it analyzed by biophysicist W. C. Levengood in his Michigan lab. Dr. Levengood found magnetic particles similar to others he has studied from England and Canada. Dr. Levengood concluded the formation soil had "remarkably uniform distributions of what appear to be iron micrometeorite material taken within and immediately outside the circle formation."

Dr. Levengood reported there was a "significant linear correlation in the relationship between the amount of magnetite material in the soil and the distance from the observed epicenters of the sampled circles. This type of linear distribution was shown to agree with the fundamental physics of centrifugal forces on particles suspended in a rotating plasma vortex system."

That means the amount of magnetic material in the soil increased toward the circle perimeters.

It's true that meteoritic iron particles are found in soils all over the planet and might explain these anomalous findings independent of the formation's creation in the oats. But the uniform linear distribution of the iron consistent with "centrifugal forces on particles suspended in a rotating plasma vortex system" is not explained by meteors moving through the atmosphere, and again raises the question: What is the energy source for the hypothesized rotating plasma vortex?

Nancy Talbott also pointed out there was a difference in the iron distribution in two of the five circles. In two, the oats were laid out straight from the center point in a radial lay, not spiraled.

"What we found," Nancy told me in a radio interview, "was that the spiral circles revealed a disposition of this magnetic material that fits the centrifugal mathematical predictions of amounts that would be deposited according to the laws of centrifugal force — more would be going to the perimeter of the circles. But in the radial lay, there were more particles at the center and less at the perimeter."

That is consistent with the hypothesis that radial lays are produced by energy coming straight down into a center, while spinning plasma vortices impose a spiral lay throwing particles outward. The implication is that two different energy processes were involved in the creation of the Marion, New York, oat formation in August 1997.

Iron micrometeoritic material found in Marion, New York, pictogram. Distribution correlated with the "physics of centrifugal forces on particles suspended in a rotating plasma vortex system." Photograph © 1997 by W. C. Levengood.

Back in Wiltshire, Nick Nicholson also remembered that the first anomalous photograph he took, photograph 5, "was in 1990 at a place called Swafield at White Horse Common in North Norfolk. Earlier in July 1990, I had traveled to Wiltshire and spent most of my time in the Vale of Pewsey. On my return, I visited my daughter and family in Norfolk and discovered that their area also had been visited by the circle makers. I managed to find the farm family that had two crop formations.

"The family had a strange tale to relate. The night prior to the discovery of two formations at White Horse Common, they had been preparing to go to bed. The farmer's wife had her attention drawn by a strange noise and she went to the bedroom window. Upon drawing back the curtains, she beheld a golden ball descending slowly towards a clump of trees. The golden ball was encircled by a ring of flashing lights as it disappeared behind the trees. Then a flash of lightning leaped skywards which was met by a downward flash. The following morning as harvesting started, the circles were discovered. This was after the batteries went dead on their farm vehicles.

"When I went into one of the circle formations to photograph, I did not notice anything peculiar at the time. But when my film was developed, I discovered the odd out-of-focus section in one of the frames I took of the wound-up 'nest' at the center. You can see the oddly soft-focused area in photograph 5 surrounded by clear focus. I don't know what to make of it." Could a low-energy plasma linked to the formation energies affect photons and "smear" the image?

Photograph 5 shows out-of-focus section in upper central stems of swirled crop center. Surrounding stems are in focus at White Horse Common, Swafield, North Norfolk, U.K. Photograph © 1990 by Nick Nicholson.

As Nick talked to me about the mysterious light phenomenon, we did not know until later that three small white balls of light were videotaped by Donald Fletcher of London, who saw two of the three lights moving low and rapidly above the overlapping crescents at Barbury Castle. That 265-foot-diameter formation discovered on July 23, 1999, was one of four patterns in the hill fort area since that spring. Others were the menorah and horned crescent rings found May 31, followed by the triple crescents and then the long serpent on August 5.

Unidentified lights videotaped in this wheat formation discovered on July 23, 1999, near Barbury Castle, 165 feet in diameter. Video frame © 1999 by Steve Alexander.

Mr. Fletcher told me that in the early afternoon of August 7, 1999, he was videotaping with his camcorder in the Barbury Castle crescents when he caught a glimpse of two white objects, one moving especially fast. He focused his camera on the fastest object and got about thirty seconds of its motion over the formation on tape. As the object passed low over the tramlines, it appeared to be about the size of a football. Frames 1 and 2 show the very bright unidentified object not far from people walking in the formation who do not seem to be aware of its presence. Frame 3 shows two more white objects catching up and passing it. At the video production house, as we moved the tape frame by frame, the editor and I were puzzled by the first object's motion. At normal speed, it had rapid, unwavering flight. But at slow frame-to-frame rate, the object appeared to rotate end over end.

Frame 1: White object moving rapidly across triple crescent formation at Barbury Castle, August 7, 1999. People in formation do not seem aware of moving light Donald Fletcher saw and taped. Videographs © 1999 by Donald Fletcher.

Frame 2: One and a half seconds after frame 1.

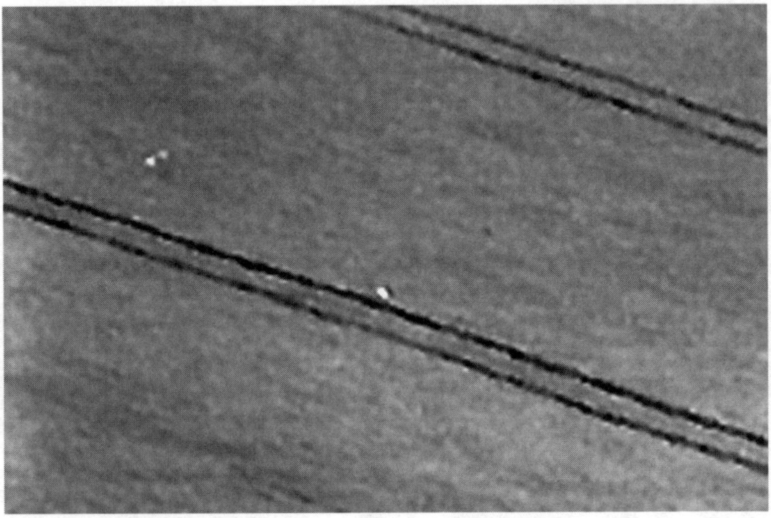

Frame 3: At center of photo is unidentified object Donald Fletcher is following with his camcorder 9.5 seconds after frame 1. The two white unidentified objects to the left catch up to and pass the central white object. Videograph © 1999 by Donald Fletcher, London, England.

Five days later in Beckhampton near Avebury, a "folded" and woven hexagram, or Seal of Solomon, was discovered in wheat on July 28, 1999. Camera and electronic malfunctions were reported in the formation. Ron Russell introduced me to musician Masao Maki from Boulder, Colorado, who had been in a group meditation with several visitors from Japan the day before the hexagram appeared. The group had focused on communicating with the force behind the crop circles and asked for a pattern that would be meaningful to their Japanese culture. When Masao saw the design at Beckhampton, he was convinced their meditation was successful."

"The circle makers responded with origami!" he said excitedly.

Origami is the Japanese art of folding paper into decorative shapes. Perhaps the point of the circle makers was that sacred geometry and its underlying mathematical principles is their language, and even a Seal of Solomon can be made with folds.

Beckhampton, Wiltshire, "Origami Hexagram," or Seal of Solomon, discovered July 28, 1999, in wheat, 72 meters (236 feet) in diameter. Aerial videograph © 1999 by Peter R. Sorensen.

The Barge had been filling up to overflow while Nick, Carl and I talked. People were coming outside along the canal where it was cooler. A man called out to us, "Did you hear about the two guys who discovered Roundway this morning? They heard loud static sounds, so loud they were afraid to go in!"

"Where are they?" I asked.

"Camped over at the Golden Swan in Wilcot."

4

Roundway and Beyond

Electrostatic crackling and other peculiar sounds, such as Charles Mallett's "heartbeat" experience in the Torus Knot, have occasionally been heard by people inside or near crop formations since at least 1989. That year, a BBC television crew accidentally recorded a noise while filming in a circle at Beckhampton in Wiltshire. The BBC audio man, Richard Merrick, compared what he heard in his earphones to "an electronic sparrow. It comes and goes." The electronic chirping sound was especially strong when he stood at the formation's center.

As early as 1990, unusual noises had been reported and audio recorded half a dozen times. Colin Andrews compared the sounds to "many, many crickets all operating at the same time. It is a noise at 5.2 kilohertz."

At 6 AM on Saturday, July 31, 1999, Alan March and Jonathan Williams walked into such intense electrostatic crackling that they stopped, afraid to go farther. The place was Roundway. In front of them was the magnificent formation of seven inner and seven outer circles. In the days before, Alan had had an intuitive hunch that something would happen at Roundway. In fact, his feeling was so strong that he left his friend, John, on a nightwatch at Avebury and drove by himself to Roundway not long after sunrise.

I wanted to hear their story firsthand and tracked them down at their camper behind the Golden Swan in Wilcot. Campers and tents are the lodging of choice for people who spend days or weeks hunting crop formations each summer. Camp space can be rented for modest weekly or monthly fees in contrast to expensive bed-and-breakfasts and hotels that charge by the night. Alan and John had traveled together to Wiltshire from their homes in Harlow, Essex, simply to get close to the crop mystery that provoked them both.

To reach the Golden Swan, I traveled a narrow road from Alton Barnes and the Barge east past Adam's Grave. In Wilcot, the Golden Swan is a lovely English country house and pub surrounded by trees, flowers and pasture. There are only four or five rooms upstairs for bed and breakfast lodgers, but outside is a large grassy field away from the main house reserved for tents and trailer campers, similar to the camping pasture at the Barge.

John and Alan had been coming to Wiltshire since 1991 to explore crop formations, had already been there several days and were ready to return to Essex when Roundway occurred. When I knocked on their trailer door on August 2, John and Alan had not been up long. They reasoned that if the six-petalled flower at Avebury and the seven-armed "throne" at Roundway both came down on Saturday morning, they had to stay up Saturday and Sunday nights to see what else might show up. But those nights had been quiet.

I explained that at the Barge I had learned about their hearing a loud noise at Roundway that reminded me of the snapping, crackling and popping sounds in Brentwood, Tennessee, and Hoeven, Holland.

I told Alan and John I reported for radio and asked if we could go to Roundway together for an interview at the spot where both had encountered the noise. They agreed.

That second trip back up the hill reminded me of my first trek down another hillside into a pictogram at Milk Hill in 1992. That's when I sensed something profound staring back at us from the unseen. Something not quite comprehensible was behind the strange pattern of circles connected by thin, curved pathways.

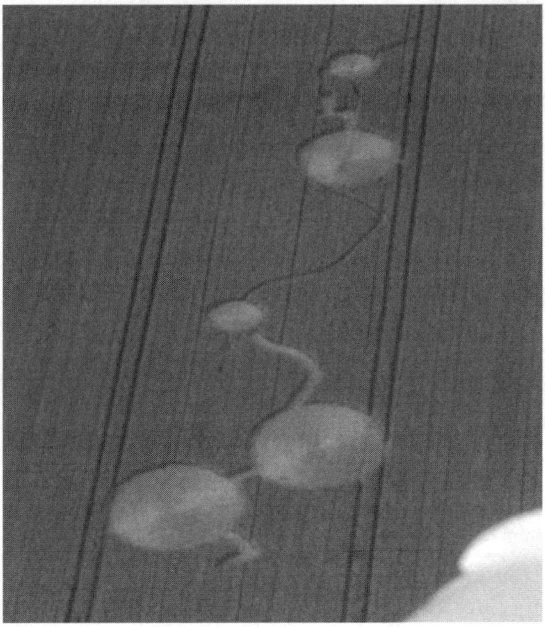

Milk Hill formation found July 16, 1992, below the White Horse, 360 feet long, in wheat. Aerial photograph © 1992 by author.

After we got out of our cars, I stopped again at the top of Roundway Hill to look at the massive and perfect imprint in the wheat. Why this sevenfold shape now? And at Roundway? The last formations there were on July 21, 1995, a "dot" at the center of a ring with a larger circle orbiting inside the ring's perimeter. (See Chapter 4.) The second was a single misshapen circle under the large power lines that crossed the field.

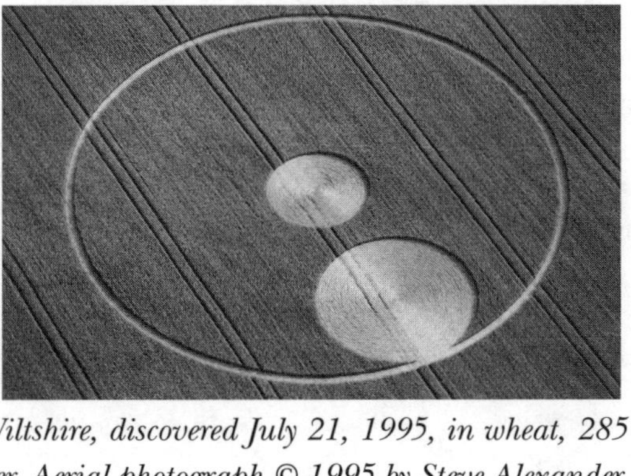

Roundway, Wiltshire, discovered July 21, 1995, in wheat, 285 feet in diameter. Aerial photograph © 1995 by Steve Alexander.

Michael Yudowitz from Berkeley, California, photographed in both formations around July 24. When his film was developed, one frame had an unidentified white object in the misshapen circle that he had not seen at the time.

Original photo above and enlargement below of unidentified object in Roundway circle formation, July 24, 1995. Photograph © 1995 by Michael Yudowitz.

Close-up of unidentified object in Roundway circle. Photograph © 1995 by Michael Yudowitz.

Alan and John went ahead into the sheep pasture and I followed down into a tramline left by the farmer's tractor. At the edge of the formation, we stopped. We were now where Alan and John had first encountered the strong noise on July 31. I ran my tape recorder as the two men explained what happened. Since Alan was the one who found Roundway, he began.

"When I arrived, this was in the field here. I had no means of photographing it properly, so I drive back up to Avebury and pick John up and come straight back here and we were first in."

"In fact," John added, "the first thing we had to do was to find a way into the formation. There were no broken paths. We had to kind of sneak into the tramline and make our way down to the formation. And once we got to the bottom of the hill, it was crackling. Pretty loud."

Alan reinforced, "It was alive! It was so fresh and we could only assume it had been done very recent to my first seeing it."

Alan March and Jonathan Williams talking with author on tape about their 6 AM encounter on July 31, 1999, with the loud electrostatic crackling noise at the Roundway formation. Photograph © 1999 by author.

"How far away were you from the formation when you first heard the sound?"

"The crackling?" John asked. "As I came down toward the formation, I became aware that it was an electric pylon sound, but kind of exaggerated over the whole formation."

Alan continued, "I was ahead of John in the tramline and didn't hear much until I got right to the edge of the formation. I was walking along and obviously excited. I wanted to get in quick and I was walking fast. But for some reason when I arrived right at the edge of the pattern, I stopped. I could hear like the whole pattern was crackling. And it was really loud and it made you feel uneasy. So I walked back up the tramline to John and said, 'Come down and have a listen.' And he came down and you couldn't hear it until you were right at the edge of the pattern."

"Were you both nervous about going into the formation?"

Alan answered, "That crackling noise stopped us going in for a while. In fact, we backed away from it and went down another tramline to photograph it from farther away for maybe twenty or thirty minutes. Then we decided to try going back in."

One of the first photographs taken between 6 and 6:30 AM on Saturday, July 31, 1999, at the edge of the fresh Roundway formation where strong electrostatic crackling sounds persisted. Photograph © 1999 by Alan March.

By then, John said that the electrostatic sound had dropped in volume by at least half and they both felt safer about entering the wheat.

Alan said, "Everything about it affected us as the freshest one we've ever been in. We've been in morning ones, but they didn't have the fresh feel of this one. I said if we don't go in while the noise was there, someone else will just go in, so we might as well go for it. The static was coming off the power lines there as well. But the static was right *in* the wheat when we first arrived."

Early morning photograph of one of the fresh coiled centers in the Roundway formation, between 6 and 6:30 AM on July 31, 1999. Photograph © by Alan March.

"When you finally decided to walk in, could you feel any heat?"

"No. But over the whole pattern, the crackling sound was really loud. And the pylon lines are way in the back, not really over the formation at all. The lines are further back and crackling like you would normally hear. But this crackling was *not* the lines! It was powerful."

"And you two have been at other formations and heard power lines?"

"Oh, yeah, we've seen other formations by power lines. We know the noise power lines make, but the loud crackling here — well, I've never experienced anything like it!"

John added, "So, we are quite confident that the noise was *inside* the formation as we entered it, this crackling noise. It filled the whole formation and going up in the air as well, it was above us as well as *in* the formation."

Photographer Peter Sorensen said he has also "been in new circles that crackle, including the East Field, and there are *no* power lines there."

"Alan, because you seemed to be the one inspired to come, is there anything about Roundway that means something to you?"

"I can't put my finger on anything like that really. I sort of see that certain times, certain nights, you do feel as though whoever is doing these patterns can tap into your conversation and sort of lead you along.

"When I drove up in the early morning, it made my heart race because I initially thought, 'It's happening once again, listening to our conversations,' which it has done before. It's nothing concrete. But you certainly can make a direct link. We'd even been talking about wildlife and I'd said I hadn't seen a deer in Wiltshire that I could recall. And when we approached the Roundway pattern in the morning, a deer came through the field and came over to us and actually jumped into our tramline and looked at us and then jumped off again. And we said, 'Boy, everything about the morning is following our conversation.'"

"Intuitively, at a gut level, do you have a sense about what's behind the formations? What's making them?"

Alan stared at the huge swirl of wheat around us. "No, I have no idea. I have an understanding that humans didn't go out there and roll that, but no other idea what it is, to be honest."

"Yeah, I feel roughly the same," John added. "My views change every time I come and look at them. I can't really say what the hell is going on, but I just know it's well worth looking into. There's a lot of work going on. There's got to be a reason."

"And humans did not make this formation at Roundway?"

"Well, Alan was there first thing. You'd think people would be packing up their vans to go off, wouldn't you? But there was no one there!"

"I suppose," Alan said, "whoever made Roundway has been making all the good ones that I've seen for years and enjoyed. It's the same hand. If it was humans — well, I'd love to meet whoever could do that. But I find it hard to believe it's a group of humans, I must say."

John agreed. "When you arrive and you look at the size of the field, you've got to find an access and there was definitely *no* access going to this formation. So we made a small track to the tramline and then made our way to the crop formation itself, which is quite a way into the field."

Alan and John were amazed that there was also a formation at Avebury the same night as Roundway and the two men had been in both places.

John said, "We were originally only a stone's throw from that Avebury one [six-petalled flower]. Avebury is where we first went. The only reason we didn't see it is that Alan headed off for Roundway and then came back immediately and got me. Otherwise, we would have gone up the Avenue of Stones, found that flower one, and we wouldn't have even found this one [Roundway]."

The night before, on Sunday, August 1, three visitors from Bristol and Northampton were in the Roundway formation photographing. It was early evening and the sun was still up. Gill and Bob Nicholas saw their friend, Gordan Stewart, sit down inside the raised ring in the middle of the sevenfold pattern. Mr. Nicholas spontaneously snapped a photograph with the flash activated. Later Gill told me, "My husband took the photo not really knowing why, as he doesn't usually take people in crop circles. Afterward, Gordan teased Bob for taking the photograph, 'You never know, I might even have a halo around me!'

"We were all amazed when we got the photographs developed and this weird arch of light went right across where Gordan was sitting. None of us saw anything unusual at the time, so what could have caused it?"

In Chapter 8, researcher John Burke compares the Gordan Stewart photograph with one he took in Kent Cliffs, New York, at a dolmen rock mound in 1994, built long ago by Native Americans for reasons as mysterious as all the stone circles and mounds in Great Britain. One of John's photographs, also taken with the flash activated, has a white arch that shows structure within it that is similar to the Stewart photograph. That internal structure could be the classical motion of charged, glowing ions spiraling upward along one of the earth's magnetic field lines.

Both John and Dr. Levengood suspect a low-energy plasma emanated from a spot of high electrostatic charge at each site, was charged by the camera flash and glowed briefly but long enough to be captured on film. If that speculation is correct, what produces localized areas of high electrostatic charge?

Flash photograph shows mysterious arch with internal structure that seems to spiral like a plasma. Not seen visibly at the time. Gordan Stewart from Earls Barton, Northampton, England, is sitting at the center of the Roundway formation on Sunday, August 1, 1999. Photograph © 1999 by Bob Nicholas.

On Tuesday, August 3, I had dinner at the Waggon and Horses pub in Beckhampton near Silbury Hill with Ron Russell, a good friend, wonderful artist and crop circle researcher from Denver, Colorado. With us were other longtime friends, Chad and Gwen Deetken from Vancouver, British Columbia, who have spent several summers in England since the crop mystery began. I was introduced to Ron's colleague, Simeon Hein, Ph.D., a social scientist from Boulder, Colorado, who had volunteered to measure electrostatic energy in some of the 1999 crop formations. It was his third time in Wiltshire.

Simeon explained that he was using a device called a Trek-520 that measures surface voltages from 0 to 2,000 volts, positive or negative. He walked into formations holding the meter at a pretty steady three feet off the ground. He was seeing static charges from several feet outside the edge of formations and at various places inside formations. He videotaped the meter readings as an archive for later analysis. He also checked the meter in normal crops without formations as a control comparison.

He had been in the Roundway formation early the second day after its arrival, and his meter battery drained down rapidly. He and Ron also had surprisingly dramatic results in three formations: the spiral arcs at Hackpen Hill, the "Sierpinski gasket" at West Overton, and the six-pointed star at Devil's Den. The Trek-520 stopped working as soon as Simeon walked into all three. New batteries had no power at all after Hackpen and West Overton. The problem was even stranger at the Devil's Den in Fyfield Down. The voltage meter, Simeon said, "froze with a reading of minus three volts for no apparent reason. The Trek remained 'frozen' for several hours afterwards. Masao Maki, the Japanese musician and author from Boulder, Colorado, had a brand new Sony digital TRV-10, which displayed a 'dirty [tape] head' warning for several minutes and then would not function."

Simeon was very puzzled about the minus-three-volts reading. He thought it might indicate a low battery, but the warning indicator never lighted up. He went back the next day to try Devil's Den again. The same failures occurred, except that the voltage meter worked normally for about one minute before freezing up. A week later, he went back to test Devil's Den a third time. Simeon found "significantly higher readings than the surrounding fields, particularly in the small circles within the main formation." There was also a "membrane" of higher voltage around the edge of the formation.

In our dinner conversation, the consensus was that the mystery was more confusing than ever and that the older ideas of meteorological phenomena or ETs were shifting into a more complicated view of other dimensional presences. Like Charles and Frances Mallett, people wondered if the emergence of more dramatic three-dimensional patterns in 1999 implied an effort to get people to consider the unseen as important, or maybe even more important than what we see around us in our everyday lives. Ron liked to do group meditations for earth healing in the crop formations, but acknowledged he had occasionally experienced conflicting energies that troubled him.

If there is a conflict of energies, does that mean a conflict between *different* intelligences behind the formations? Have humans been cast in the drama, consciously or unconsciously? Why stage a chess game in the cereal crops of this planet?

One man I met in Houston, Texas, in March 2000 was an East Indian noted for his clairvoyance in political matters. He looked at crop formation photographs I showed him from 1999 and earlier years. In silence, he began to divide the photographs up on the table. I was intrigued as I watched him place Cherhill 1993 and some of the 1999 formations such as Roundway, Cherhill and Hackpen Hill in the same group.

Then he placed his large, brown hand on top of the Cherhill pile and said, "Angry, this is angry and wants to reclaim this world. But this!" He moved his hand to the other larger pile of photographs. "This is already here and has control. There could be a great conflict over our earth. And humans, many humans, are involved. They are watching, monitoring, learning. I don't think these humans know what is going to happen. And I don't think they want any other humans to know.

"But this angry one, I think it is God himself. Or pretty close. Very high up. And what's happened to earth has got to be changed."

After dinner at the Waggon and Horses, I had a strong desire to go up Woodborough Hill in Alton Barnes for a nightwatch and asked if anyone else wanted to go. Ron, Chad and Gwen were tired, but Simeon had never been to Woodborough and wanted to experience the famous hill at the heart of some of the most active crop circle fields in Wiltshire. Woodborough stands out for miles around as the only high hill with a stand of trees on top. These beech trees harbor legendary tales of lights, ghostly images and a palpable atmosphere of another presence I have felt each time I've been there. Below, cereal crops surround Woodborough in the huge expanses of the East Field and Tawsmead Copse.

After parking at the hill's base, we began to climb in the dark. The grass is thick, there aren't many holes to trip in and the pure darkness is such a novel contrast to city life that for a few minutes the night climb was entertaining. Then Simeon turned on his fluorescent flashlight.

When we got to the top, a couple we didn't know had a video camera mounted on a tripod. The man asked us if we knew what the blue sparkling light was in the north. We could clearly see what looked like a child's Fourth of July sparkler, but there was a regular, rhythmic pulse to the sparks. The color was a combination of sky blue and white with a regular cycle of brighter bursts. I looked at it through my Nikon zoom lens. I tried to photograph, but nothing came out on later prints. Simeon looked through his camera, too, but could not see any structure and had no idea what was producing the light. The four of us watched the blue-white "sparkler" for about forty-five minutes. Then it suddenly was not there anymore.

Woodborough Hill topped by beech trees rises above the huge surrounding wheat fields in Alton Barnes, Wiltshire, England. Photograph © 1995 by author.

A few minutes later, a pale yellow, gibbous moon rose in the east. The beauty of the surrounding fields in the moonrise was mesmerizing—until another light popped in the sky. It was sky blue, completely round, about the size of my little finger's tip held at arm's length. The blue light moved in a slight arc east to west over about ten degrees and popped right back out again. Vanished. With a camcorder and two 35mm cameras among us, nothing of that remarkable sky activity ended up on film or tape.

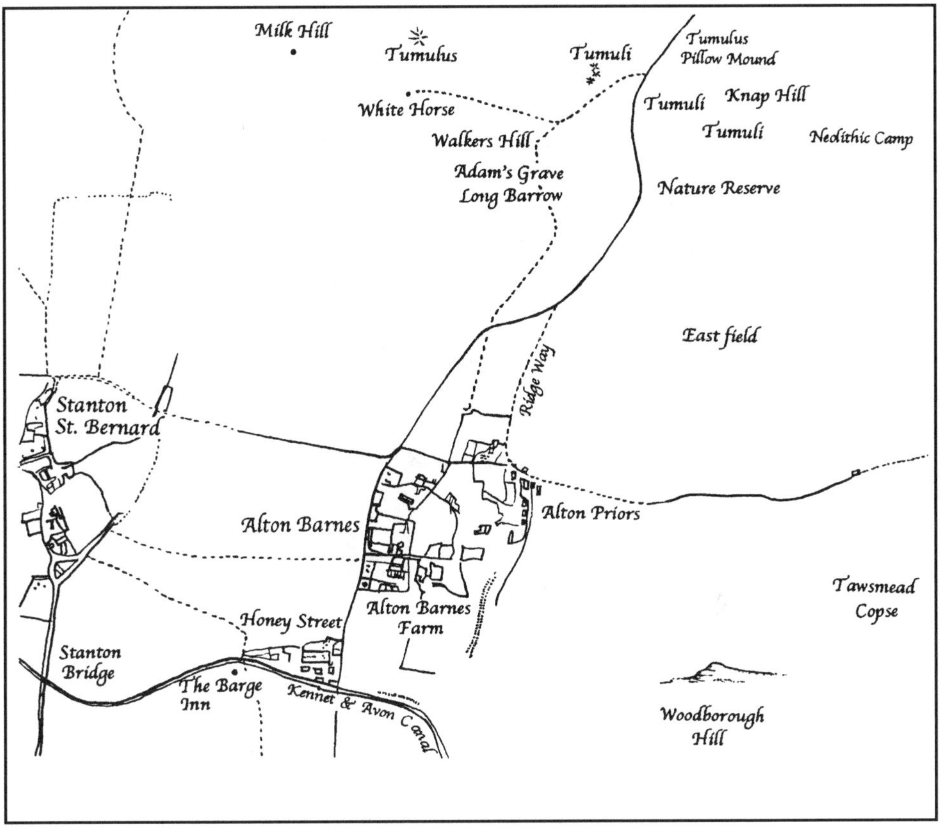

Map of Alton Barnes and Woodborough Hill area.
Photograph © 2000 by author.

The next morning, August 4, I was meeting with Andreas Mueller at his campsite next to the Barge to talk about his saros idea in the Hackpen Hill formation and to study his elegant and meticulous surveys. (See Chapter 9.)

In astronomy, the saros cycle is the interval of 18 years, 11 1/3 days after which the Earth, Sun and Moon return to nearly the same relative positions and the cycle of lunar and solar eclipses begins to repeat itself. Each saros cycle contains about 43 solar and 28 lunar eclipses over a period of 800 to 1,200 years.

Andreas had been reading a German astronomy book about the August 11, 1999, solar eclipse in Europe when he discovered a graphic of saros cycles drawn on a polar projection of the Northern Hemisphere that reminded him of the Hackpen Hill formation.

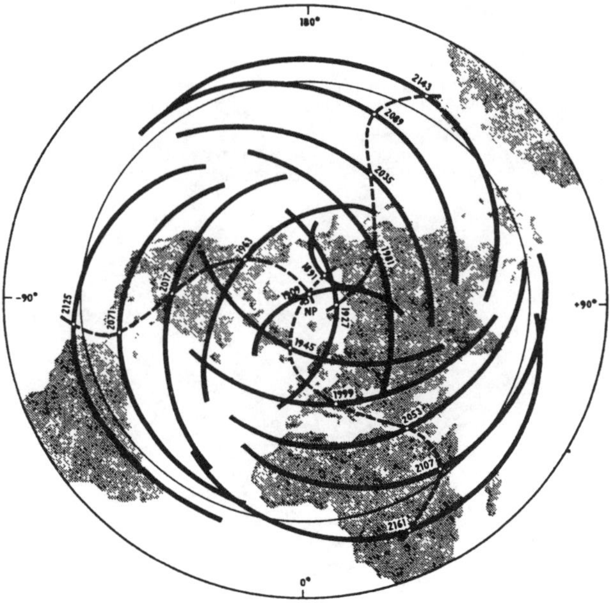

Graphic depiction of a saros cycle of solar eclipses on the surface of the earth, from Die Sonnenfinsternis am 11. August 1999, *by Wolfgang Held. Diagram © by J. Schulz.*

Hackpen Hill formation in wheat, discovered July 4, 1999. Aerial photograph © 1999 by Lucy Pringle.

Andreas later wrote: "The diagram shows the actual cycle of the 'zones of totality' wandering southward by starting at the North Pole and completes at the South Pole. This actual saros cycle is known as saros cycle #145 and had its first eclipse in 1639 and will be completed in 2648 with its last eclipse over the South Pole before a new saros cycle starts over again at the North Pole. But we should consider that the formation is not the diagram itself, but could be an artistic interpretation. It makes no difference that the diagram shows a different rotation of the spinning system because this is only a question of projection. If the earth would be shown from behind, the spinning direction would fit."

While we were going over the saros cycle graphic and his surveys, a cell phone rang. It was Ulrich Kox, who had been flying over the West Kennett Long Barrow and found a brand new formation. Andreas, his colleagues and I took off in a car caravan for the Long Barrow across the road from Silbury Hill.

The West Kennett Long Barrow is a mound 330 feet long, one of the longest in the British Isles. Archaeologists speculate it was built around 3500 B.C. I had been in the entrance tunnel only the morning before wondering about the ancient intelligence that had constructed long barrows all over England and other parts of the world. Were they only burial chambers? Or could they serve other purposes that we have not yet discovered? Andreas would later tell us that the length of the new formation was 100 meters, or 328 feet, almost exactly the length of the West Kennett Long Barrow.

Not only had I been at the Long Barrow the day before, I had tried to walk from there west toward the "ziggurat" formation across the road from Silbury. I had struggled through one large unplowed field and had started into the next wheat field when I realized that the ziggurat was much too far to reach that way. So I had turned back to the Long Barrow and the road in front of Silbury Hill.

Now, this new formation we were about to investigate was in the same wheat field where I had stopped. And I began to think about those odd blue lights the night before directly north of Woodborough Hill. Had those lights been over the West Kennett Long Barrow?

As soon as we reached the downed wheat, we were impressed by its enormous size. Andreas immediately began working with his tape measure and sketch pad.

Andreas Mueller from Saarbruecken, Germany, surveying the five pyramids formation discovered August 4, 1999, at the West Kennett Long Barrow across the road from Silbury Hill. Photograph © 1999 by author.

Andreas Mueller sketching from his tape measurements after walking the perimeter of the five-pyramid formation in wheat on August 4, 1999. Longest length was 100 meters (328 feet). Photograph © 1999 by author.

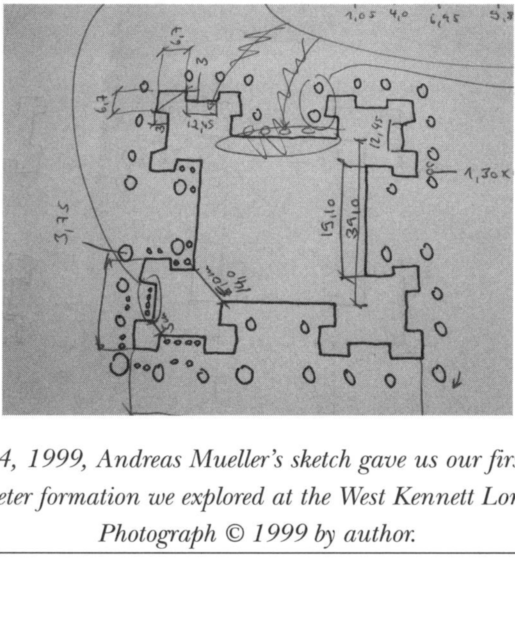

On August 4, 1999, Andreas Mueller's sketch gave us our first "aerial" of the 100-meter formation we explored at the West Kennett Long Barrow. Photograph © 1999 by author.

I walked throughout the formation thinking that the lay of the crop was especially confusing. Later, we would all be surprised when we saw an aerial photograph. Andreas's sketch looked like the foundation of a castle. But the aerial showed five "pyramids," or Buddhist stupas, rising upward. We marvelled about a design in which plants would have to be laid in very specific angles to catch light for the three-dimensional illusion.

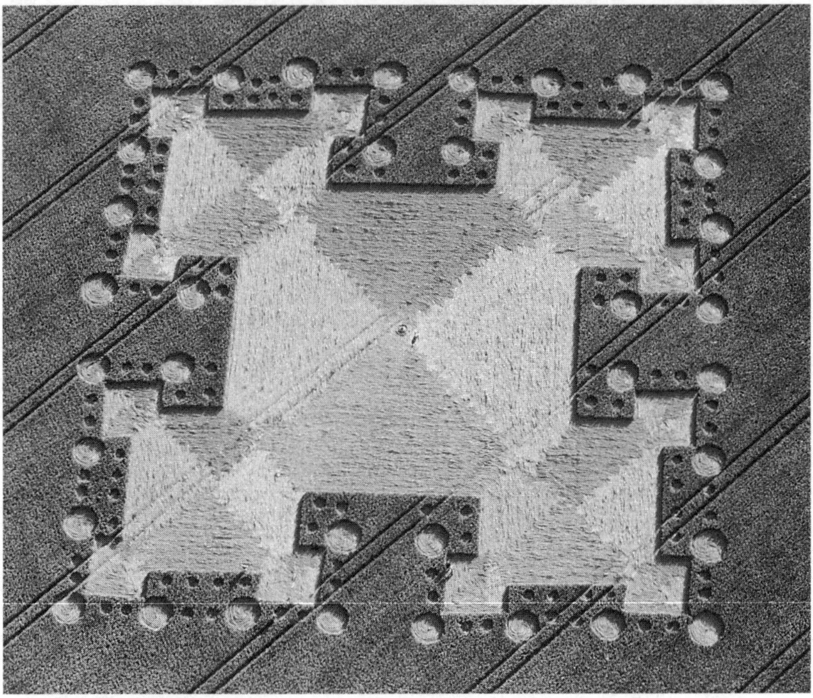

West Kennett Long Barrow five "pyramids," or Buddhist stupas, discovered August 4, 1999, in wheat, 100 meters (328.08 feet) in longest direction. Aerial photograph © 1999 by Steve Alexander.

When Andreas's survey for this West Kennett Long Barrow formation is compared to the July 16, 1999, formation at Windmill Hill, both have similar interior designs. But unlike the five pyramids, the Windmill Hill formation was inside a circle and has four avenues defined by the small circles projecting in four directions from the interior pattern. Both formations were exactly 100 meters, or 328.08 feet, in their longest length and diameter.

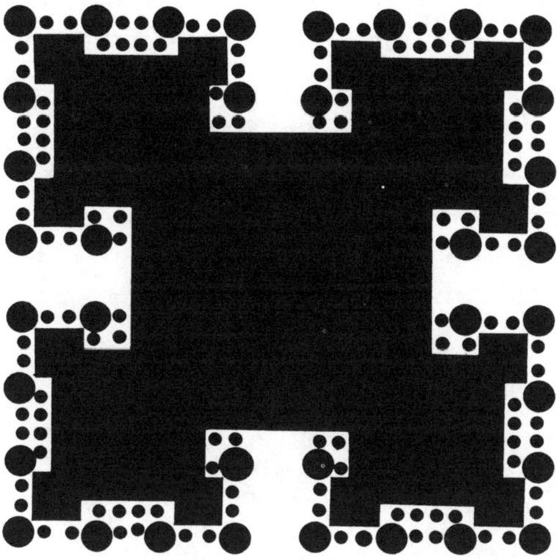

Diagram of "five Pyramids," or Buddhist stupas, at West Kennett Long Barrow, discovered August 4, 1999 in wheat, 100 meters (328.08 feet) longest length. Black interior matches the white interior of Windmill Hill formation below. Diagram © 1999 by Andreas Mueller.

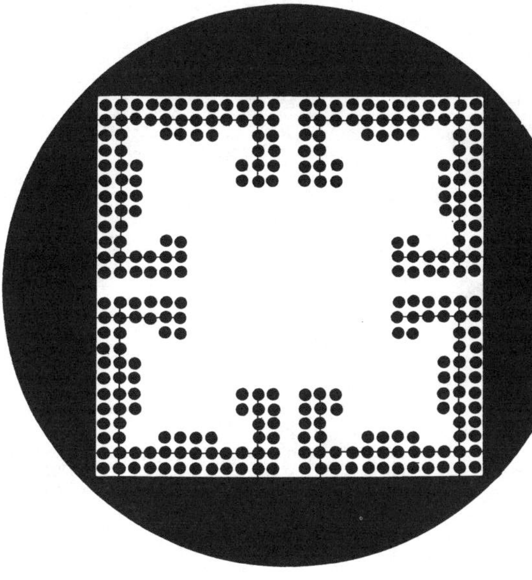

Diagram of Windmill Hill, Wiltshire, formation discovered July 16, 1999, in wheat, 100 meters (328.08 feet) in diameter. Diagram © 1999 by Andreas Mueller.

I found another interesting anomaly standing straight up from the center of one of the many circles that bordered the "pyramids." It was a green thistle blooming with bright purple flowers. The crop in the small circle had been twisted around the thistle's base. What plasma vortex could twist the wheat, but leave the thistle seemingly untouched? Over the years researchers have noticed that plants other than wheat, barley, oats or other grains are not flattened. Do the energies that create formations focus specifically on certain molecules?

Thistle plant standing in middle of twisted wheat center, August 4, 1999, at West Kennett Long Barrow formation. Photograph © 1999 by author.

I ran into Chad and Gwen later at the Barge. I told them about the strange, sparkling blue light that Simeon and I, along with the couple videotaping, had seen from Woodborough Hill on Tuesday night, August 3. I wondered if the blue light had anything to do with the creation of the West Kennett Long Barrow formation.

"You've got to be kidding!" Chad said, laughing. "Gwen and I watched a blue light, too. I was feeling sort of down that Gwen and I were too tired to go with you and Simeon up to Woodborough Hill. So I'd gone out on the balcony of the house we were in on the third floor in Alton Barnes. That way I could at least look at the sky. And as I stepped out, on my left in the direction of Devizes due west, I saw a flashing, strobing light. Initially, I thought it was a police car because it looked exactly the same — very rapid flashing and bluish-white. It was below the horizon hills, so I thought it was along the road and police had pulled someone over.

"I looked at it with binoculars and I couldn't see any background. Usually with a police car light, you see the trees lit up in the back slightly or the top of the police car.

"I kept looking away at the stars and then back at the light for about ten minutes and it just stayed there strobing. Finally, it seemed to move above the horizon. At that moment, I realized this is no police car! This is something very unique. Through my binoculars, I could see it wasn't rising straight up. It was zigzagging like a sewing machine would go back and forth. And it pulsed very rapidly. Gwen and I counted the number of flashes in a minute and it turned out to be about 2.5 flashes per second. Then it moved gradually east towards Avebury, still zigzagging. I could see it clearly for a good thirty to forty-five minutes. Then it gradually faded away and disappeared."

"Gwen, what did you see?"

"The same thing as Chad. With the naked eye, you could just see the pulsing. But when you looked at it through binoculars, you could definitely see that it was like a round ball, a perfectly round ball. And there were a lot of other craft that night. There were Hercules planes flying around on maneuvers from RAF Lyndham. There were a lot of helicopters and planes that night as well. You could tell the helicopters and planes through the binoculars because you could see the red and white flashing lights. But this strange strobing light was not a normal airplane, for sure."

"And the color you saw with your eyes?"

"Blue-white. Sort of a pulse of blue and then white and blue and white. But because it was pulsing so quickly, the blue and white kind of blended into one."

"And this was after we left the Waggon and Horses dinner?"

"Yeah."

"So you must have been watching about the same time that Simeon and I were up on Woodborough Hill watching the blue-white sparkler object that seemed to be north of us — I thought towards West Kennett."

Chad thought for a moment and said, "Well, from where we were standing on the balcony, we were facing north, too. I heard other people who saw it also. One couple watched it from the Honeystreet crop formation behind the Barge. We've tried to triangulate their position with ours and it appears to be about a mile from where we were staying and about two miles from where you were on top of Woodborough Hill."

Gwen said they got up the morning of August 4 and drove all over where they thought the light had been, looking for a formation, but could not find any. At the time, neither knew about the West Kennett Long Barrow pyramids.

Chad said, "We expected the formation to be where the light rose out of the field. But that new formation was exactly where it had *disappeared* going toward Avebury."

Chad and Gwen Deetken from Vancouver, British Columbia, in Alton Barnes, Wiltshire, the summer of 1999. Photograph © Chad Deetken.

5,000 year old Silbury Hill rises 130 feet high across the road from the "ziggurat" wheat formation discovered the early morning of July 24, 1999 after a man reported seeing the field glowing. Photograph © 1999 by Steve Alexander.

"Sentence" (1,020 feet long) and "Serpent" (500 feet long) discovered together at sunrise on June 12, 1999. Aerial photographs © 1999 by Lucy Pringle.

Unexplained anomalies in June 14, 1999 aerial photograph above, ground shot below, in East Field "Sentence" © 1999 by Francine Blake.

Strange grey "mist" was not visible when this photograph was taken on the afternoon of August 24, 1997 in Marion, New York oats formation © 1997 by Larry G. Thomas.

Strange grey-white "mist" not visible at center of 1995 Roundway formation © 1995 by Denni Clarke, Freedom, California.

Flash photographs show mysterious arches with internal structure that seem to spiral like a plasma. Not seen visibly either at Roundway formation above on Sunday, August 1, 1999, photograph © 1999 by Bob Nicholas, or at Kent Cliffs, New York rock dolmen below © 1994 by John Burke whose electrostatic voltmeter needle froze on maximum positive scale.

August 11, 1999, Stonehenge, Lara Stefancin-Nelson does not see three translucent spheres which also appeared in other people's photographs that evening © 1999 by Lara Stefancin-Nelson.

July 28, 1999, Hackpen Hill, video frame of orange-red light that pulsed and moved for several minutes © 1999 by Mark Haywood and Carl Nevin.

August 22, 1992, camera aimed toward Avebury stones. Translucent sphere not seen at time by photographer, Nancy Talbott, who says no flash was used for afternoon shot after thunderstorm © 1992 Nancy Talbott.

Above: August 8, 1999, odd sphere with black "holes." Below: Sphere with singular black hole. Neither was visible to people attending meditation at Martinsell Hill, Wiltshire. Photographs © 1999 by Ed and Kris Sherwood.

July 24, 1995, Roundway, Wiltshire circle formation, unidentified object not seen at time. Color slide © 1995 by Michael Yudowitz.

June 19, 1999, East Field, Wiltshire "Sentence" formation on an overcast morning, light and blue spot below it not seen at time. Color slide © 1999 by Frank Laumen.

July 25, 1994, Wilsford, Wiltshire ring formation, moving light on videotape not seen about 1:15 PM on a clear afternoon. Light first moved down toward formation ring, disappeared, reappeared, moved down again and then back up over a period of 2.5 seconds.
Videograph © 1994 by Foeke Kootje and Conny de Bruyn.

August 7, 1999, Barbury Castle three crescents formation, moving light on videotape was one of two rapidly moving objects seen by photographer at time.
Videograph © 1999 by Donald Fletcher.

Whitefish, Montana formation oat plant with severe somatic developmental abnormality. Photograph © 1999 by W. C. Levengood.

Avebury, Wiltshire formation wheat plant with similar somatic developmental abnormality. Photograph © 1999 by W. C. Levengood.

Epidermal cell structure of Cherhill wheat leaf imprinted in thin magnetic iron glaze. Photomicrograph magnification 40x © 1993 by W. C. Levengood

Tiny glassy beads of unknown origin have been found in soils or surrounding area of crop formations in Holland, Israel and the U. S. Some are hollow and some solid such as these 1.9 to 3.7 micrometer diameter spheres found near Logan, Utah formation in April 1999. Composition largely oxides of aluminum, silica, calcium and titanium.

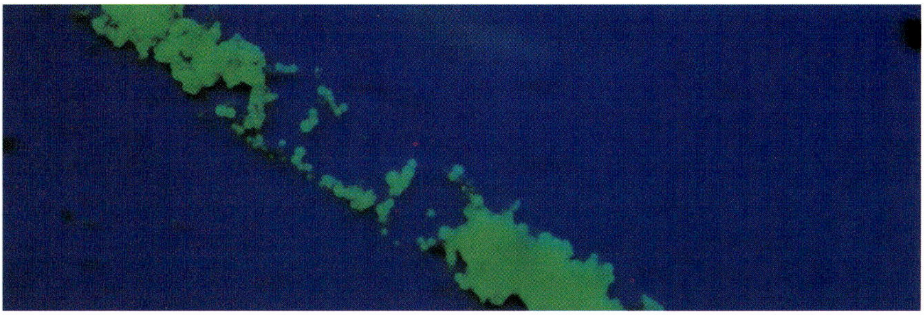

When glassy beads exposed to ultra violet light, they fluoresced a bright green. Both photomicrographs at 450x © 1999 by W. C. Levengood.

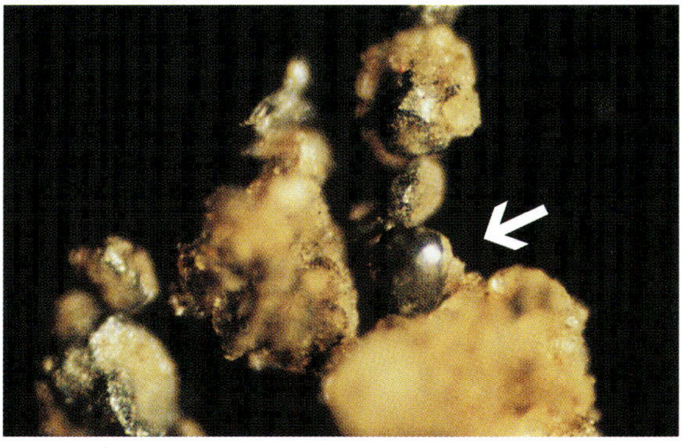

Shiny magnetite bead amid soil particles collected from August 1998 Saskatchewan, Canada wheat formation.
Photograph © 1999 by W. C. Levengood.

Accelerated growth from English formation seeds compared to normal controls. Photograph © 1991 by W. C. Levengood.

Delayed growth in yellow crescent and tiny circle compared to rest of field that had already bloomed. Aerial videograph © by Peter R. Sorensen.

Madera, California grass from formation on right still vigorous after light and water deprivation compared to normal control on left. Photograph © 1999 by W. C. Levengood.

Milk Hill, Wiltshire formation discovered May 2, 1999 in yellow flowering oilseed rape, 250 feet long. Charles Mallett encountered sphere of light inside the circle at the far right. Photograph © 1999 by Steve Alexander.

Burghasungen near Kassel, Germany formation in oilseed rape discovered May 1, 2000 after strange blackness descended twice in the night seen by twenty people. Photograph © 2000 by Frank Laumen.

West Kennett Long Barrow after mysterious transformation on June 3, 2000 from first phase on June 2nd. Aerial videograph © 2000 by Peter Sorensen.

Cherhill Down, Wiltshire, July 18, 1999 with "Signature."
Photograph © 1999 by Steve Alexander.

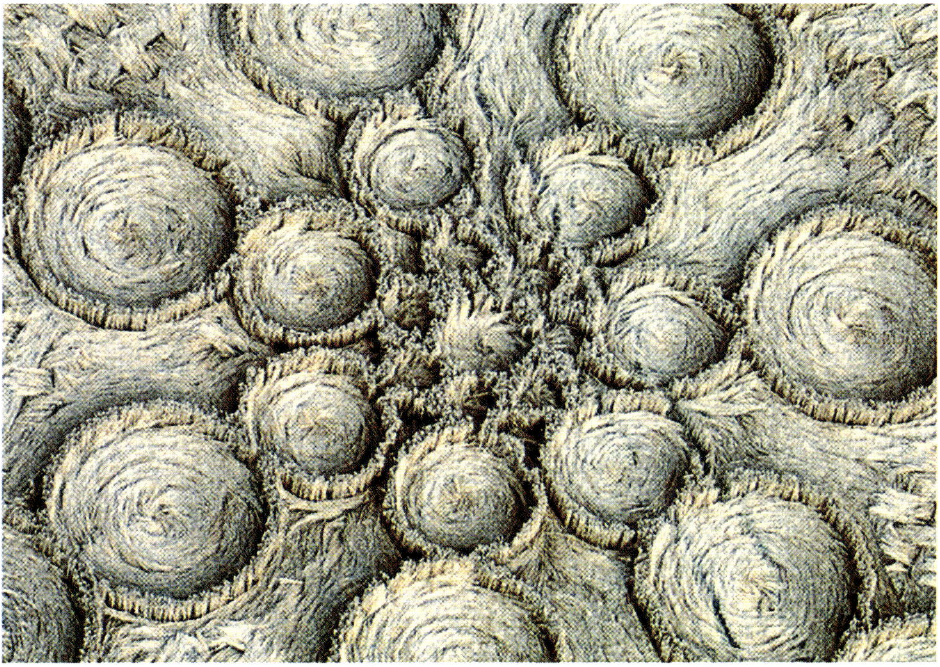

Aerial close-up of "Basket" center discovered August 6, 1999 at Bishops Cannings near Roundway. Aerial videograph taken by Ulrich Kox and © 1999 by Ulrich Kox and Peter R. Sorensen.

Milk Hill White Horse in Wiltshire above a double Von Koch Snowflake fractal that was discovered in wheat on August 8, 1997, 264 feet diameter and 198 circles. Aerial photograph © 1997 by Lucy Pringle.

The end of that week, on Friday, August 6, 1999, Andreas Mueller and another crop circle researcher from Switzerland, Werner Anderhub, were "driving around as we do every morning, looking for new crop circles, and we found the formation now known as 'the Basket' at 6:15 AM."

"Basket" wheat formation discovered in the early morning of August 6, 1999, at Bishops Cannings before the farmer erased it entirely with his tractor against all pleas to temporarily save it; 46.3 meters (152 feet) in diameter. Aerial videograph taken by Ulrich Kox and © 1999 by Ulrich Kox and Peter R. Sorensen.

At the same time, Ulrich Kox from Niederrhein, Germany, decided it would be a good, clear morning to fly and look for new formations in the Pewsey Vale. Ulrich is a photographer who has traveled to England since 1994 to shoot aerial photographs of the crop formations. In 1999, he frequently went up in a microlight plane with his three still cameras. He also had a cooperative agreement with Peter Sorensen to shoot Peter's videotape camera. Ulrich knew a man from Paris named Wilfrid Houdouin who was visiting Wiltshire for the first time and wanted a chance to fly and photograph. Together they drove to a local airfield and Ulrich arranged two flights. He would fly with his pilot to Roundway near Devizes with all his cameras, and Houdouin with his own camera and pilot would follow.

Ulrich told me, "I was flying along the Pewsey Vale between Beckhampton and Devizes at the A361 opposite West End Farm around 8:15 AM when I saw this astonishing pattern below in a wheat field. The ground seemed to be woven and even the edge was an artistic work. It looked like a basket standing in a crop field."

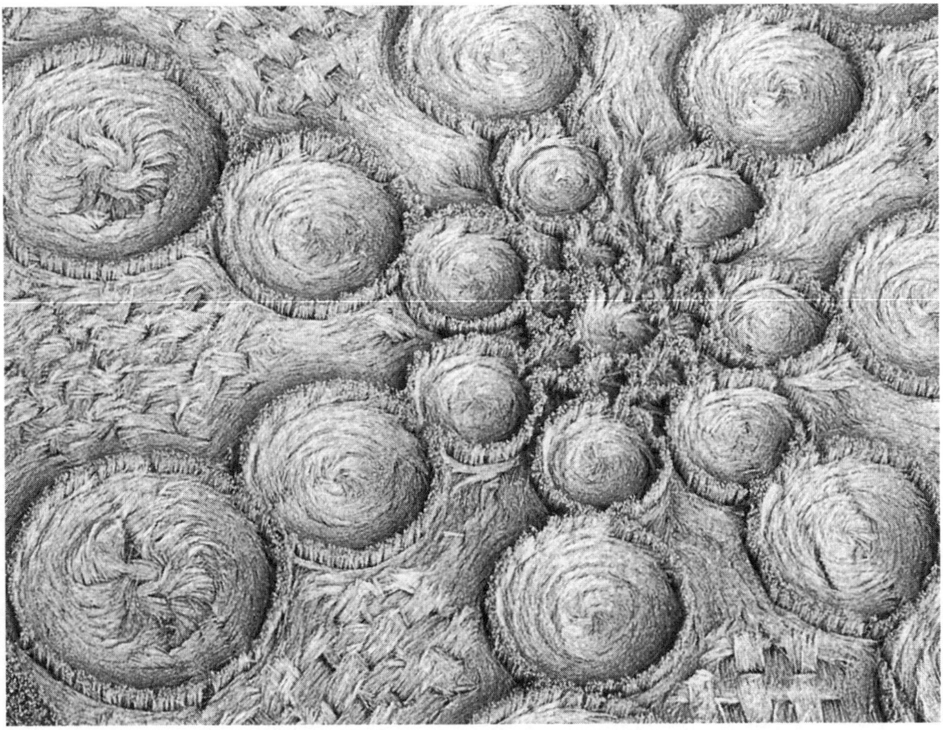

Aerial close-up on center of "basket" discovered August 6, 1999, at Bishops Cannings over the hill from the Roundway formation. Aerial videograph taken by Ulrich Kox and © 1999 by Ulrich Kox and Peter R. Sorensen.

Fortunately, Ulrich flew that morning because he ended up with the only professional, clear aerial photographs and videotape of a formation so intricately woven in three-dimensional form that it is considered by some to be unique in the history of crop formations. What happened on the ground that made Ulrich's photographs unique is described in a joint account by Andreas Mueller and Werner Anderhub for Andreas's International Crop Circle Archive (ICCA) and also posted at www.cropcircleconnector.com, an excellent crossroad for crop formation photographs and eyewitness and research reports on the Internet created and edited by Mark Fussell and Stuart Dike in England.

"As far as we could see, the formation was untouched. The lay was very soft and interesting, especially those areas that created the basket-style which were flattened in a completely new way as we have never seen before in a formation. Other beautiful features were those thin, standing rings that reminded us of the famous 'Missing Earth' formation at Longwood Warren in 1995. The Basket's rings were placed on underlying paths that started just at the first sets of rings to continue to the very centre of the outer rings, creating spokes.

"Basket" at 6:15 AM on August 6, 1999, at Bishops Cannings as Werner Anderhub and Andreas Mueller approached. Photograph © 1999 by Werner Anderhub.

"Some of the flattened circles inside the standing rings had only one stalk standing in its centre while some others were nicely twisted around them. The very central circle offered a special detail — or was it only coincidence that it showed exactly *seven* stalks still standing?

Seven wheat stalks standing in the center swirl where the seven spokes in the basket weave came together. Photograph © 1999 by Werner Anderhub.

"As we could already see from the ground, the sevenfold geometry was done very well and we got confirmed by the later aerials taken by Ulrich Kox and Frenchman Wilfrid Houdouin. We also surveyed the formation. To us it was very obvious that nobody was inside before us because the flattened areas showed no crushing marks nor muddy footprints. The night before was very humid. Even though we walked with care, we left tracks and crush marks.

"As we arrived back at the campsite later that morning, we heard that the farmer had indeed cut out this beautiful formation. Frank Laumen, Bert Janssen and others filmed the action at the scene where they were able to discuss with the farmer. However, this discussion was more a monologue by the investigators. To nearly each request of the investigators, the farmer answered with, 'It's just vandalism.'

"Some investigators offered the farmer money to stop the harvester engine and wait just another ten minutes so they could have a short look at the formation that was, at the time, not reached yet by the harvester. But he denied. He strangely even refused when Dutch researcher, Bert Janssen, offered him 100 English pounds (about $160).

"The farmer repeated, 'No, it's just vandalism. If you want to see the formations, then ask those people who do them to do them in your country. Not here.'

"This same farmer in 1994 would not give public access to see the famous 'Devizes Scorpion.' And in 1998, the same farmer cut out a fresh formation before aerials were taken. So, we had the strong feeling that his actions would continue in 1999."

Andreas Mueller and Werner Anderhub managed to make a detailed survey of the extraordinary formation (see Chapter 9) before the angry farmer arrived. Later, Werner wrote further in a field report for the cropcircleconnector web site:

"As you are all aware now, the farmer harvested this wonderful and incredibly important formation the moment he found out about it. These images are the only images that are available (taken before the farmer cut the formation out).

"This formation, I believe, is an important shift for the phenomenon in 1999. Never before have we seen such a detailed floor construction as we now have *raised* woven crop to create the basket effect.

Raised woven section amid standing rings. Photograph © 1999 by Werner Anderhub.

"Its appearance is highly impressive from the aerial images taken. But not only that, the circles themselves also have some fascinating rotations in them. If we look closely at the detailed aerial images, we can clearly see that these swirls are different to what we have seen before.

"Each of the largest circles placed on the outside of each arm has also a weave rotation to them. If you look at the circle on the far right, with the arm positioned at four o'clock, we can see a swirl more detailed than the rest, with a raised centre creating a two directional 'S' swirl. The circles further down the arm do not seem to display this feature. They are formed from a more conventional rotation. But the smaller circles in the centre are far too small to comment on. Indeed, all of the circles on the outside have this 'S' swirl, but what is so staggering is the raised basket weave in between the arms."

I left for France the next day, August 7, to join friends for the solar eclipse watch on August 11. Back in Wiltshire on August 9, a young woman named Lara Stefancin-Nelson — seven months pregnant and visiting crop circle country for the first time — went to the "ziggurat" formation across the road from Silbury Hill. She asked a friend to photograph her so later her child-to-be could see "how blessed he was before he was born."

Lara Stefancin-Nelson from South Lake Tahoe, California, on August 9, 1999, happy to be in the "ziggurat" across the road from Silbury Hill, her first crop formation in England. Translucent sphere left of center. Photograph © 1999 by Lara Stefancin-Nelson.

She did not know until her photographs were developed that a few contained translucent spheres, including one during her first moments kneeling down in the formation. The translucent sphere is a mysterious anomaly that has appeared in crop formation photographs throughout the 1990s. The next day, on August 10, Lara traveled with a group to Hampshire and the East Meon farm country southeast of Andover, where many formations have occurred outside Wiltshire over the years, including three in 1999 that had a similar quality and design.

At Henwood near East Meon, Hampshire, a formation was discovered on August 3 that was similar to the West Meon and Warnford patterns. Lara and her traveling companions were eager to see the relatively fresh site beyond Wiltshire.

East End near West Meon, Hampshire, barley formation, discovered June 21, 1999, 130 feet in diameter. Aerial photograph © 1999 by Lucy Pringle.

Lippen Lane, Warnford, between East and West Meon, Hampshire, wheat formation, discovered July 12, 1999.
Aerial videograph © 1999 by Peter R. Sorensen.

Henwood, near East Meon, Hampshire, discovered in wheat, August 3, 1999. Aerial videograph © 1999 by Peter R. Sorensen.

Large pink glow at left and white object at center were not seen in Henwood wheat formation on August 10, 1999, near East Meon, Hampshire. Photograph © 1999 by Lara Stefancin-Nelson.

White object appears to move from left to right almost like a time lapse, but no light was seen in the Henwood formation on August 10, 1999, near East Meon, Hampshire. Photograph © 1999 by Lara Stefancin-Nelson.

Over the past several years, Lara has photographed for musicians in well-known rock bands and considers her 35mm camera very dependable and trouble-free. She shot ten or more rolls of film in England without any camera problems.

So she was surprised later back in California when she found odd light anomalies and strange spheres in about twenty-five prints out of more than two hundred. Mysterious light anomalies were in three of the photographs she took inside the East Meon formation on August 10. Some will argue these must be pressure marks in the emulsion from developing or other mechanical or chemical processing problems. Or damaged film. But her photo shop said the negatives were all right and had no explanation for the light images.

I met Lara for the first time at the Barge August 14 before she returned to her California home and just after I returned from France. Her photographs would not be developed for another month. She was sitting with Nick Nicholson and longtime crop circle researcher ilyes from the United States, who prefers the one name without a capital. I met ilyes back in 1992 on my first trip to Wiltshire to see the crop formations. She invited me to join her and a small group to celebrate her August 11 birthday at Stonehenge. ilyes has made annual birthday trips with friends and visitors to Stonehenge, where lunar and solar eclipses were tracked for thousands of years.

ilyes applies for permission from the British government to take her birthday groups into the interior of the great sarsen stones. So in 1992, with deep gratitude to ilyes, I was able to touch the huge sculpted rocks. Beyond their beauty, I felt a deep sadness, as if I sensed Stonehenge when it was new, surrounded by deep green grass on a younger, cleaner earth.

Since the totality path in England's Cornwall was cloudy during the eclipse, ilyes and Nick asked me what had happened in France. I told them it was amazing luck that my friends and I had agreed long before to meet in Reims for the solar eclipse. On August 11, satellite weather reports had focused on northern France as having the best chance of clear viewing. More than half a million people drove to Reims to watch what happened.

Reims was once the territory of the long-haired Celtic clan of Gaul. By 817, a Carolingian cathedral was begun and over centuries became the now famous Notre Dame Cathedrale of Reims. A Reims high school was located where the treaty ending World War II was signed.

I traveled with astronomers from the Reims Planetarium to view the eclipse while my friends stayed at the cathedral. Andy and Vittoria Abercrombie from Houston, and Derick Rawson from Galveston, Texas, had a digital still camera. That way, we could increase our chances of get ting at least some images of totality. I took off to the country with astronomer Sylvain Dardenne and a dozen of his astronomer colleagues. Most of them had never seen a total solar eclipse before, either. That's why they wanted to get away from the city and up on an unobstructed hill in order to see the first shadows of totality rush toward us.

We were about twenty miles north of Reims trying to stay dry inside tents on the night of August 10. We had champagne and wine and homemade quiche to celebrate being there in spite of the cold downpour outside. By sunup, the sky was low and gray, but the astronomers persisted with setting up several telescopes and cameras all equipped with the correct filters to view and photograph the sun. Hundreds more people arrived at the hill, and the roads were lined with cars. Cell phones became the walkie-talkies for everyone to check on the Reims sky. Heavy clouds were still reported there while our hill-top clouds at least kept breaking open as the first black bite of shadow on the sun began at 11:06 AM French time.

Astronomers on hilltop twenty miles north of Reims, France, August 11, 1999, as total solar eclipse began at 11:06 AM after all-night rain. Photograph © 1999 by author.

Cameras clicked through thick filters as the hour and eighteen minutes ticked down to totality. Then, as we all thought we were going to see the beautiful terror of darkness at noon, a thicker cloud moved to block our view. There were groans and then screams as a blue hole opened up behind us in the valley. We all began to run toward an open field in the direction of the only sunlight. Five of the astronomers decided at the last second to get into the only four-wheel drive there, and took off across the muddy stubble wheat fields toward the sunlight a couple of kilometers away. We all wanted to be in that car, but what happened next stopped me in my tracks.

The very white clouds and blue patch of sky over the small area of gold light in the village about a mile away began to transform into the strangest dark blue tinged with green and gold that I have ever seen. As if someone had turned down a rheostat, the whole sky darkened quickly, so quickly it was scary. Astronomers compare the color of total solar eclipse to the deep twilight forty minutes after sunset. But from that French hillside — perhaps because of all the moisture — the sky was an eerie, deep royal blue misted by green and gold in which at least one bright star emerged. Everyone became quiet. The air moved in the sudden temperature change. Birds in the wheat stubble began to fly low in zigzag patterns as if disoriented.

Then we heard the screams of the men in the four-wheel below us in the village. They had made it to a place beyond the clouds where they could clearly see the eclipse. When they returned with a camera broken in the wild ride, they said their violent dash was worth it because "never have we seen such light," Sylvain said. He was surprised that the sun's face did not turn black. "It was royal blue with green and gold, a color on another planet!" He was grinning like a child as he spoke rapidly in French-accented English learned during studies in the United States. "And when the last full ray of the sun was going out in the 'diamond ring,' the corona began to expand right and left glowing a more strange blue, not white!"

About 12:27 PM, August 11, 1999, just after totality as the sun's crescent emerged beyond the moon north of Reims, France. There will not be another total solar eclipse in England until 2090. Photograph © 1999 by author.

I saw this diamond ring phenomenon when I finally looked away from the drama of the wheat stubble and village light and up behind me into the sky where the sun should have been. There was a spark of light as the first sun rays moved through the cratered terrain of the moon and then flooded the surrounding clouds, changing them to an abalone shell, swirling and glistening in greens, golds, blues, pinks, white.

Back in Reims, Andy Abercrombie photographed the crowds holding filtered glasses over their eyes as the solar eclipse was projected on a large outdoor video screen. As noon light dimmed to night, all the lights in town came on.

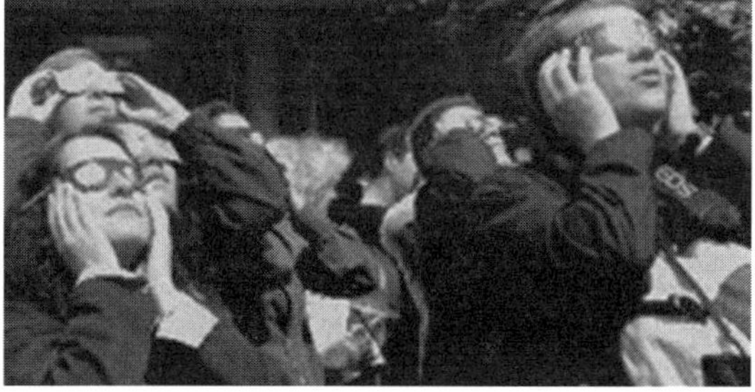

Reims sun-gazers held filtered glasses to their eyes as the August 11, 1999, solar eclipse progressed toward totality. Photograph © 1999 by Andy Abercrombie.

Streetlights in Reims, France, go on as the August 11, 1999, total solar eclipse turn noon to night. Photograph © 1999 by Andy Abercrombie.

Video screen in front of Reims Cathedral shows progression of the August 11, 1999, total solar eclipse. Photograph © 1999 by Andy Abercrombie.

During that same hour near East Looe, Cornwall, England, crop circle researchers Ed and Kris Sherwood had finished a group meditation with friends on Bindown golf course.

Ed and Kris said, "From our location, the sun was obscured by clouds for most of the eclipse, including the moment of totality, so we were looking in a westerly direction to witness the approaching shadow and darkness. As the moment of maximum darkness began, a bright, stationary, reddish light appeared above the horizon to the southwest of our position about five miles away near Polean Farm and the village of Pelynt.

"Using a High 8 video camera, Kris began taping the luminosity. Despite darkness and overcast, we could see the light clearly. It was brighter than any other object in view at the time. Then as Kris attempted to zoom in on the first object with the camera momentarily out of focus, four more reddish lights appeared. Three lights were in a line with one below, perhaps a mile to the right of the first light. They looked identical and we considered that they might be parachute flares. But we did not see smoke trails, nor could we see a floating, slow descent you normally associate with flares.

"Before Kris could refocus on the group of now five lights, the first luminosity disappeared after maintaining a stationary position for approximately twenty-five seconds. The four lights then remained in view for almost a minute after the first before they, too, all vanished together.

"Later, when we viewed the tape a frame at a time, we noticed that the four lights appear to flicker or pulsate a little, especially the luminosity in the top right corner of the group. When the tape is in slow motion, that light seems to disappear and reappear often. The sighting lasted approximately one minute twenty seconds, of which we recorded about one minute five seconds.

"Reviewing our experience and footage, we concluded that the reddish lights were not airplanes, helicopters, balloons or radio mast lights, as there weren't any at the sighting area at the time. Though we cannot be totally sure they were not parachute flares, I question whether the military would have released large flares above public land, close to an ancient sacred site and a village at the moment of total solar eclipse.

"For Kris and me, the luminosities appeared to be either what earth mysteries researchers call 'earthlights' or 'atmospheric plasmas' sometimes reported as luminous UFOs. Or it was an elaborate hoax that involved parachute flares."

Videotape showing orange lights that suddenly appeared near northern horizon at solar eclipse totality on August 11, 1999, in Looe, Cornwall, England. Video frame © 1999 by Ed and Kris Sherwood.

The August 12, 1999, edition of the *Cornish Guardian* reported, "Some sky watchers [of the solar eclipse] in Cornwall say they saw things yesterday that they were not looking for — UFOs.

"There were many reports of hovering balls of light being spotted during the period of total darkness, which *shot off in a southerly direction* as soon as light returned [Howe's emphasis]. Whole families and complete strangers standing next to each other claim to have witnessed the phenomena which was also captured on video."

This first newspaper report that the lights "shot off in a southerly direction" is not consistent with flares that burn up in one spot.

But by August 19, the *Cornish Guardian* offered an official explanation for the mysterious lights under the headline of this article supplied by Kris and Ed Sherwood from Dave Gillham, founder of the Cornwall UFO Research Group.

Flares spark reports of UFOs.

Pranksters are suspected of sparking a UFO scare during last week's total eclipse by launching parachute flares into the darkness. Thousands of sky watchers reported seeing three or four orange-colored lights apparently hovering in the sky for about a minute during the solar spectacle on

> Wednesday morning.
>
> The *Cornish Guardian* received a flood of telephone calls shortly afterwards from people trying to solve the mystery of what they had just seen.
>
> And on the morning after the eclipse, when we published a report of the sightings, a further 50-plus calls came in from readers, two of whom had captured the strange images on videotape.
>
> But we have established that the lights were parachute flares, probably launched from handheld canisters somewhere between Par Docks, Carloggas Downs and Fowey. PC Pete Keene, training officer for Launceston police, who was watching the eclipse from high ground at Blisland said, "They were definitely parachute flares. I was able to see them fairly close up, but I can understand how people looking from a distance might have thought they were something else."
>
> At the request of the *Cornish Guardian*, both videotapes showing the lights were scrutinized by Terry Hearn who operates a video editing suite in Roche and who has over 15 years experience of electronic imaging. "I carefully examined the tapes frame by frame and am 100 percent certain they show flares," he said. "In fact, I can even detect the smoke trails."

Ed Sherwood is not so certain that the official explanation is correct, because he has seen mysterious lights for years that are not military flares. He told me, "When I came to Wiltshire for the first time in 1992, very quickly I began to see light forms which I had not seen before. They seemed to be invisible to most other people and I began to think of them as nonphysical emanations from another dimension or frequency. Then, on July 26, 1992, I witnessed nonphysical light forms *transform* into a visible and physical light over a wood, move across the wood and down into a wheat field, and when we got to the location soon afterward, there was a new crop circle. So I know from my experience that the lights have physical connections to the crop circles."

"Why did you go to Wiltshire?"

"Basically, I was living up in Derbyshire more than a hundred miles away. And one day, I had a very strong clairvoyant impression that I must go down to Wiltshire, particularly to a location I didn't know before. I looked at it on the map and it was a small wood near Alton Barnes called Tawsmead Copse. I had this very strong impression I had to go there because there was going to be a major interaction experience with the genuine circle-making source and people. I've had these kinds of impressions all through my life and I've learned to listen to them.

"So I jumped on a train the following day and went straight to Wiltshire and put my tent right next to the trees at Tawsmead Copse. I learned that at least forty other people were coming to nearby Woodborough Hill on July 22 for experiments by Steven Greer, M.D., for his Center for the Study of Extraterrestrial Intelligence (CSETI). Some who came included Ron Russell, Shari Adamiak, George Wingfield, Maria and John Ward, Michael Hesemann, Lynn Gladwyn, Reg Presley, Colin Andrews and many others.

"A few hours before the large gathering, I was standing alone in a wheat field. I'd had an ability for many years to clairvoyantly sense the appearance of physical energy balls, or what scientists would call 'atmospheric plasmas.' I asked the Infinite Intelligence and Consciousness of God as a 'clairvoyant medium' to help in the group's effort to initiate interaction with the 'Source' or 'Sources' of crop circle formations that were not made by humans.

"Hours later around dusk, I was on Woodborough Hill with the group about to arrive. I suddenly sensed an infinite 'presence' above me. I looked up, and like an answer to my request for help, saw for the first time a fully materialized, three-dimensional-looking, pearly white, non-physical ball of light. It glowed clearly visible for about three seconds before vanishing. During those seconds, I could see it was about a foot in diameter and perhaps ten to fifteen feet away. Colin Andrews and Maria Ward were standing next to me, but they did not see it. Within minutes after that, I began to see a flickering, brilliant white light in my peripheral vision that would suddenly appear, draw my attention, then disappear as soon as I tried to look at it.

"Reappearing every few minutes or so, this light was different from anything I had experienced before. It took a few times of turning my head to look in vain before I fully realized the source of the peripheral 'light' was external from my own consciousness. Suspecting it was the 'Creator' as the 'Cosmic Joker,' manifesting psychokinetically to attract my attention like a poltergeist that only I could see at the time, I challenged the Source inwardly and thought, 'OK, you're playing with me, but if you can't appear in the center of my vision, I'm not going to keep turning my head to look.' The psycho-interactive response was immediate. The very next luminosity appeared in the center of my vision. I thanked the Source and from that moment to the present day, I have seen nonphysical luminosities every day, mostly this way.

"During the second night of the CSETI experiment, I began to see nonphysical light balls multiply and cluster over hills, trees and water sources. I noticed that 'clustering' light orbs would occur rapidly or slowly, last for a few seconds to several minutes and, depending on weather conditions, could transform into physical luminosities which I think are atmospheric plasmas or 'earth lights.' These were visible to everyone present in an ordinary unaltered state of consciousness. Reg Presley filmed an 'orange' plasma from Woodborough Hill after I saw nonphysical light balls, invisible to the group, cluster at the same location.

"Four nights later, on July 26, 1992, near West Stowell in Wiltshire, I was observing nonphysical light orbs cluster above a small wood when they suddenly disappeared and were replaced by a large ball of orange light instead. Approximately twenty to thirty feet in diameter and visible to at least ten other people in two different locations, it remained stationary for about two minutes before descending into a wheat field where a new ringed crop circle formation was found the following morning. I think that ultrasound frequencies are used to resonate formation plants, 'tuning' them to stand, laydown, spiral, or whatever direction each plant is to go."

"Ed, what's the difference for you between extraterrestrial vehicles that might have ETs in them and the intelligence or force associated with the light spheres?"

"Well, for me the difference is that the intelligence behind the physical energy balls which some call earth lights, atmospheric plasmas, or dragons — depending on the culture — is what I would call the Infinite Intelligence. It's the intelligence of creation. Call it God, if you like, but definitely an intelligence that is infinite.

"With the extraterrestrial vehicles, I always sense before I even see them that one is about to appear at a location in the landscape. I get a very strong feeling of beings with individualized consciousness, personality, ego, identity, individual life-forms.

"But the intelligence that I experience with these nonphysical luminosities and the physical luminosities that can transform from the nonphysical is not a personality, but something infinite — something that can take on many forms and can easily reflect your own psyche. But intrinsically, it is beyond form. It can take on a matter form on this planet, perhaps other planets throughout the universe, according to natural law.

"For example, what I've learned over the years studying crop circles is that I noticed there seemed to be a collective of cosmic and earth forces, processes and conditions that bring about genuine crop circle creation. For example, the physical balls of lights, the earth lights or plasmas — I've seen more than a hundred, including this summer of 1999. They will form under certain weather conditions and not others, as if for them to occur at all there have to be conducive conditions, according to natural law."

"What kind of physical conditions do you think are necessary for the spheres of light to appear?"

"For the physical spheres of light that people can see, in terms of meteorological conditions, you need little to no wind. Plasma, which is a ball of ionized or electrified gasses, is bound by a powerful magnetic field. But if you have a strong gusting wind that will disperse that energy, the magnetic field of a plasma cannot overwhelm the more physical and powerful wind. So I have never seen plasma forming under high wind conditions. I've also never seen it form in rain. I've seen it form either side of rain. High humidity is very important. There can be fog, but not large raindrops falling. The energies being electrical and particularly electrostatic plasma, if it comes in contact with liquid water, it loses its charge and will be absorbed by the water."

"Ed, why would an infinite consciousness need to interface at all through water molecules in sphere shape?"

"Yeah, it's a good question. What I'm trying to say is that certain laws or forces were set in motion at the beginning of this universe. What I've learned from research and experience is that the Infinite Intelligence, the Collective Intelligence, within all of creation — and not just the creation of this universe, but all universes before it and all after — this Unquantifiable Intelligence manifests according to natural law. After that is set into motion, the Infinite Intelligence does not go against that.

"For example, if the force of gravity on this planet is of a certain strength, the Infinite Intelligence does not say, 'Ah, today, I think I'll make it a hundred times stronger.' That might not be a problem for the Infinite Intelligence, but it would be a tremendous problem for us.

"What I've learned is, we have a situation where we come from an Infinite Source and what we call our soul is infinite in its essence. It's timeless and spaceless and is not from a dimension of physicality with physical laws and forces.

"When you hear the different religions speaking about individuals achieving enlightenment, superconscious awareness and God consciousness, Nirvana, different names — basically, it is the consciousness of a mortal being reaching out to the Infinite. As much as we are trying to reach out and understand this tremendous Infinite Creation, the Infinite is trying to reach back. And as we know throughout life, there is duality: hot and cold, day and night, male, female. One of the most fundamental dualities of existence and creation is the finite and the infinite. The infinite by itself is not God and the finite by itself is not God, but the two together are.

"We have photographs of nonphysical light balls that people don't generally see, but they show up on photographs. One type we've only recorded during rain showers. That's not a contradiction. The *visible and physical* light balls I've just talked about can be absorbed by water. The *non*-visible and nonphysical ones operate differently.

"A good example is one I photographed on top of Martinsell Hill on August 8, three days before the solar eclipse. It was taken a few minutes after a ley-line activation and planet healing meditation. It shows a single nonphysical light ball with what looks like black holes in it.

Martinsell Hill near Pewsey, Wiltshire, August 8, 1999. "Nonphysical light ball" with odd black "holes" not seen at time of photograph "after a short ley-line activation meditation." Photograph © 1999 by Ed and Kris Sherwood.

"Why did it have holes? Perhaps because the rain was impacting upon our consciousness, puncturing our concentration, the Source reflected back those punctures psychokinetically."

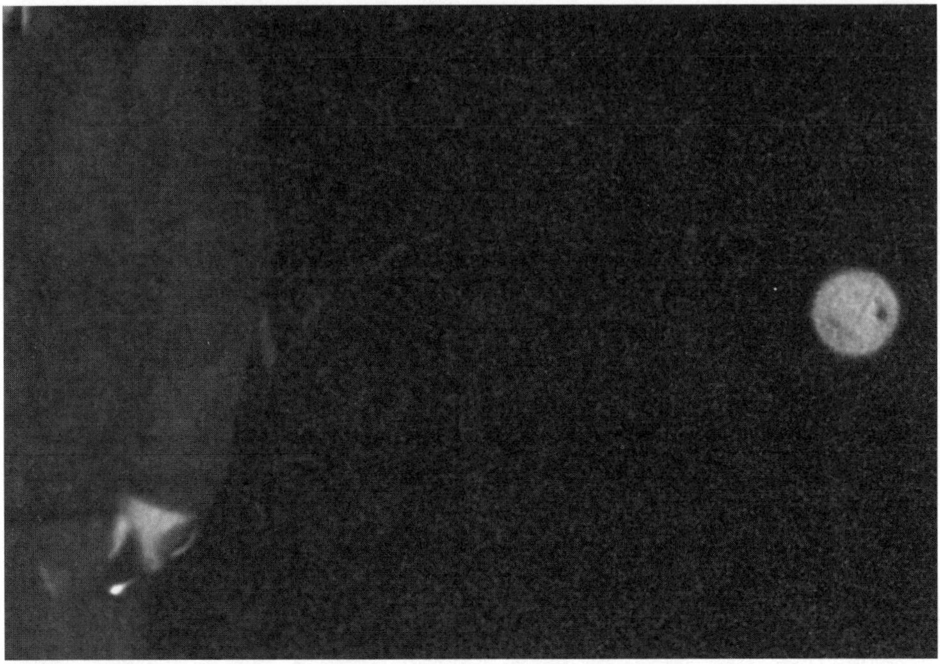

August 8, 1999, another light ball with black "hole" not seen at the time appeared in this Martinsell Hill flash photograph. Photograph © 1999 by Ed and Kris Sherwood.

Back at the Barge Inn in Alton Barnes, Lara Stefancin-Nelson told me about joining ilyes's birthday group at Stonehenge. The English weather had been too cloudy for them to see the sun clearly, but her real joy, later in the early evening, had been to walk inside and around the Neolithic stone rings.

Again, Lara asked a friend to photograph her with the great sarsen stones that have stood in circles for at least five-thousand years to keep track of solar, lunar and other cosmic cycles. More translucent spheres showed up in two of her photographs. In one, she had spread her arms out to enter the stone circle, and three translucent spheres are visible: one at the center of two stones ahead of her, one at her back and another to her left near tennis shoes kicked off on the grass.

Stonehenge, west of Andover, on August 11, 1999, Lara Stefancin-Nelson's first visit to the sacred site with ilyes and her birthday group after the solar eclipse. No translucent spheres were visible at the time, but three appeared on this photograph and one on another. Photograph © 1999 by Lara Stefancin-Nelson.

Another one of the guests at ilyes's birthday party was Janet Ossebaard from Gieten, the Netherlands. Janet has investigated crop formations for several years and coproduced with Bert Janssen their second documentary in 1999, entitled *Crop Circles, The Research.* Janet is a professional photographer, was using her trusted 35mm camera and was amazed when she got her film developed after her 1999 stay in England to find several sphere anomalies she did not see at the time she was photographing.

One of Janet's slides shows a man sitting in front of Stonehenge columns. To the man's left (page right) is a large translucent sphere that no one saw at the time, but which showed up on Janet's slide as the other spheres appeared in Lara's photographs.

Translucent sphere at right between sarsen stones photographed on August 11, 1999, during ilyes's birthday party at Stonehenge in the early evening after the total solar eclipse. Photograph © 1999 by Janet Ossebaard.

Blow-up and enhancement of sphere.
Photograph © 1999 by Janet Ossebaard.

In yet another photograph that Janet took, of a woman touching one of the Stonehenge columns, three spheres were discovered on the slide. Bert Janssen also enhanced that photograph so the spheres could be seen more clearly. The top sphere next to the arrow is nearly invisible.

Janet said, "It was about 8:15 PM when our group entered Stonehenge. Earlier that day it was closed because of the eclipse. Officials were afraid of riots. I used a flash to photograph on slide film. Later, when I got the film developed, on two of my slides are spheres — big and nearly transparent. It's really an anomaly.

"And one slide even contains *three* of them. One of our friends is standing in front of the stones and you can see a big sphere right at the head, a big sphere in the air and a third one coming out of the ground. Two-thirds of the ball is out and one-third is still at the bottom of the slide. I said, 'Wow, what is this?'"

So Janet and Lara were at the same place at the same time, but with different cameras, and still the strange spheres ended up on their two different film formats of prints and slides, the prints developed in California and the slides developed in the Netherlands.

Three translucent spheres on one slide, August 11, 1999, at Stonehenge. Photograph © 1999 by Janet Ossebaard.

There are mundane explanations for some photographic anomalies, such as emulsion defects, roller pressure in developing and lens flares at the time photographs were taken. Some camera experts also argue that spheres and lights on photographs are a product of camera flashes interacting with water molecules in the atmosphere. The day and evening at Stonehenge had been cloudy, but not rainy. Janet Ossebaard used a flash to light her night scenes. Lara Stefancin-Nelson's camera had flash capability, but she is not certain if it was activated for the early evening photo that shows the three spheres.

For others who did not see lights at the time and yet light images showed up on prints, Nick Nicholson took daylight photographs in 1995 at Goodworth Clatford, in 1991 at Tawsmead Copse, and in 1990 at Swafield. The early August afternoon that Michael Yudowitz photographed in the 1995 Roundway circle below the power lines was clear and sunny. In all four of these cases, there was no rain and no camera flashes. Lara took clear afternoon photographs at Goodworth Clatford in 1999, and neither of us used a flash in the bright afternoon sun at Liddington Castle, discussed in the next chapter. But there had been an earlier thunderstorm. Larry Thomas in Marion, New York, photographed the odd, grey "mist" on a bright afternoon in 1997. No rain, no flash. When Bob Nicholas took Gordan Stewart's photograph sitting in the center of the 1999 Roundway formation, he used a flash. But the August 1 night was hot and dry.

For other photographs in which lights were seen at the time, such as Dr. Joachim Koch's photo at Woodborough Hill in 1999 and Donald Fletcher's videotape at the Barbury Castle triple crescents, the afternoons were clear, so neither rain nor camera flashes were involved. The July 1999 evening on which Charles Mallett photographed the orange light near Avebury was hot and dry.

5
Vesica Piscis

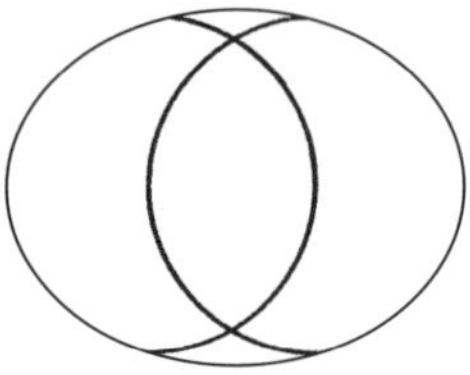

Before I had gone to the Barge, I learned that at least one formation appeared on the day or night of the solar eclipse. Four large rings were found in the early morning of August 12 near Chiseldon by German photographer Ulrich Kox, while he was flying in a microlight. The middle two rings, he said, overlapped like a Vesica Piscis symbol. At the Barge, ilyes knew where the rings were in Chiseldon. She had been there, and thought they were quite elegant and not an Audi car commercial, as some of the sarcastic pundits had teased.

ilyes started giving me directions about how to get to Chiseldon on the southern outskirts of Swindon. The key, she said, was a maize corn field not far from the Chiseldon turnoff. The next wheat field beyond the maize would have the four rings.

Lara spoke up, "Could I go with you? It will be my last chance in England to get into a crop formation before I go back to California tomorrow."

So seven months pregnant Lara Stefancin-Nelson and I took off in my rented car to find the four rings. I took the scenic route from the Barge back through Avebury instead of the more direct A345 and headed north on A4361 toward Swindon and Chiseldon. When we reached Avebury, the thunder was loud and rain was coming down hard. I asked Lara if she wanted to keep going since the rain might not stop for a while.

She laughed and said, "I'd rather be outside in the rain going to find crop circles!"

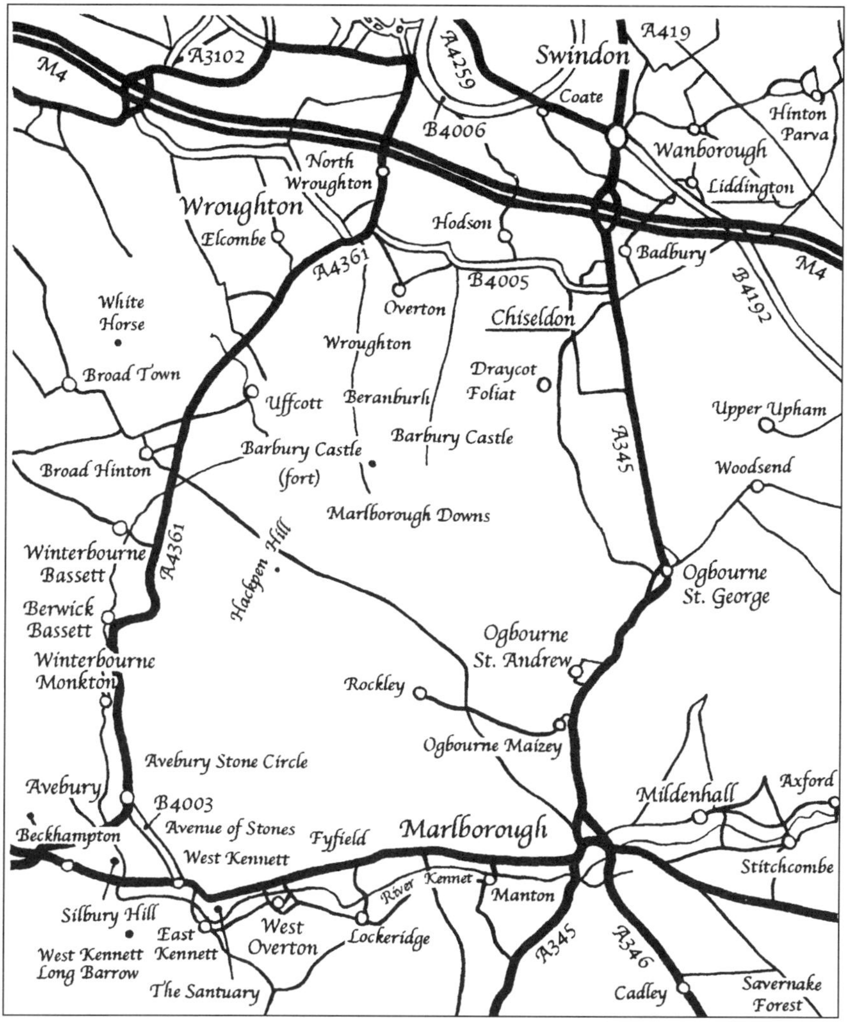

Chiseldon and Liddington in upper right quadrant.
Map © 2000 by author.

Sometimes I wonder if innocent exploration simply for the joy of it is given encouragement by the universe. When we turned at the sign for Chiseldon and started looking for the maize field, we couldn't find it. But the rain stopped, the sun started to shine again and I simply drove looking for hills and more elevation that could give us a view of the surrounding fields. Neither one of us knew exactly where we were when I looked to my left toward a beige-gold hill covered with wheat. I could see a crop formation in it.

"Lara, look at that!"

"But it doesn't look like four circles."

"No, but let's go see. It's either been there for a while or it's new. If it's new, we have to take a sketch back to the Barge and get a pilot out here for an aerial."

We turned around, went back to the main road and crossed over a major intersection. Our only guide was to keep heading toward the hill. We found a country road running alongside the field, so it was easy to park on the shoulder grass and walk a wheat tramline. Lara and I both had our 35mm cameras, and when we got into the large formation, I told her that the crop looked too old and frayed for something new. But the pattern was so precise and the lay so complex that whatever it was, I was impressed.

August 14, 1999, complicated lay inside Liddington Castle wheat formation twenty-five days after first discovery on July 21, 1999. Photograph © 1999 by author.

I began to sketch what I could figure out on the ground until I heard Lara yell, "Linda, look! A double rainbow!"

Storm clouds had blown away, and directly above us in a blue sky were one strong rainbow and another fainter one above it. Lara snapped a photograph of the rainbows and me standing in the outer ring of the formation.

August 14, 1999, after rainstorm, a double rainbow over Liddington Castle, Wiltshire, England, formation with author standing in outer ring. Photograph by Lara Stefancin-Nelson for author © 1999.

Then Lara walked toward me with her camera and took another photograph at closer range. The air was a dry summer breeze by then and neither of us used any flash. Back in California, when Lara got her roll developed, an odd anomaly resembling some of the balls of light phenomena could be seen in the lower left corner of the frame. Neither of us saw any unusual lights at the time.

The arc of the strong rainbow lingered for some time, and I took photographs looking up the hill from the formation. When I got my prints developed in Philadelphia, I noticed a bright point of light above the trees. Then, in another photograph taken later when the sun was lower, a bright point of light was still in the air above the trees.

We later learned at the Barge that this was the famous Liddington Castle pattern first discovered on July 21, 1999. I had not been to Liddington before. So, in the way the universe sometimes teaches to let resistance be your guide and let doors behind you open up, we went to find four rings we could not at first find and ended up in a formation I had wanted to see before France after a huge rainstorm that quickly changed to sun and a double rainbow.

August 14, 1999, another anomaly in lower left corner not seen at the time. Photograph © 1999 by Lara Stefancin-Nelson.

Bright point of light above trees at Liddington Castle wheat formation on August 14, 1999, and possibly same light in later afternoon photograph on next page. Photograph © 1999 by author.

Light in upper left corner similar to light above trees in earlier photograph at Liddington Castle formation, August 14, 1999. Photograph © 1999 by author.

Liddington Castle soon after its discovery on July 21, 1999. "Signature" in upper left corner. Aerial videograph © 1999 by Peter R. Sorensen.

Pole shot of Liddington Castle formation. Photograph © 1999 by Peter R. Sorensen.

The sun was getting low in the west when we finally left the beautiful formation. When we reached the car I said, "Lara, I want to try one more time to find those four rings, OK?"

"Yes! Let's drive back to that first road where we were looking for the corn field. I didn't see any maize, but maybe we didn't drive far enough."

I got to the road and moved about 10 mph along the crops, looking for the green leaves of corn that would mark the next wheat field as the place to enter. But Lara saw the rings first in the distance about half a mile from the road, where a wheat field extended up and over a hill.

I turned into a rough tractor path to park. As I reached for the door handle to get out, I saw, only a few feet away, two long, brown ears poking up out of the wheat with two enormous brown eyes below. The ripe crop covered everything else, giving the impression that the upper half of the deer's head was simply floating on the seed heads. Lara and I reached for our cameras, making signals to stay quiet. We both started moving for a better angle when the deer jumped straight up and bounded off with another deer hidden nearby.

Two deer running in wheat field on August 14, 1999, below the four rings at Chiseldon, Wiltshire. Photograph © 1999 by Lara Stefancin-Nelson.

The sun seemed to be setting fast, and I told Lara I would run ahead to try to photograph the four rings before the light was gone. Soon, I realized I was going uphill at a steeper rate than the field looked to the eye from a distance. In fact, when I got to the top, I was surprised that the upper circle went over the top of the hill and down several yards on the other side, reminding me of Charles Mallett's observation about the 1997 Torus Knot lapping over the hilltop at Alton Priors.

The width of the rings was narrow, only about six feet. Many plants stood single file along the midline of the ring paths as if unaffected by whatever force had put the rings down.

Rings were about six feet wide and long stretches of wheat plants stood single file at the middle of the ring paths. Photograph © 1999 by author.

Walking up tramline on August 14, 1999, near sunset, toward four new rings discovered in the early morning of August 12, 1999, by Ulrich Kox, not far from Liddington Castle. Photograph © 1999 by author.

When I finally saw Peter Sorensen's aerial videograph of the rings, the Vesica Piscis pattern in the middle was clear. This was the central diagram of Sacred Geometry for Christian mysticism. It is constructed by drawing two equal circles so that the center of each lies on the circumference of the other. In his *Sacred Geometry,* Robert Lawlor wrote, "The fish-shaped central area is one source of the symbolic reference to Christ as a fish. Christ symbolically connects heaven and earth, above and below, creator and creation,... a deeper penetration of spirit into form where the Word becomes flesh." This Vesica Piscis symbol of overlapping rings also lies on top of the Chalice Well at Glastonbury in the garden of the Red Spring, about an hour west of Avebury. The longest-running annual crop circle conference is held in Glastonbury each summer.

Chiseldon, Wiltshire, four rings about 250 feet long with two grapeshot at bottom center, discovered by Ulrich Kox on August 12, 1999. Vesica Piscis in middle. Aerial videograph © 1999 by Peter Sorensen.

The legend of Glastonbury is that metal merchant Joseph of Arimathea went to the counties of Cornwall and Somerset two thousand years ago to buy tin from the Mendip mines, not far from the Tor. With him was the young Jesus Christ, son of Joseph's niece, Mary. While there, Christ and the others are said to have drunk water from an open spring at the base of the Tor known today as the Red Spring. An underground well issues water through a pipe into a manmade shallow pool also shaped like the Vesica Piscis symbol to honor Christ. The flow is said to have been strong and steady for at least two thousand years. Modern measurements are 25,000 gallons a day of water impregnated with iron. The water has an iron metallic taste, but I have drunk a lot of it without harm.

Entrance to the Chalice Well in Glastonbury, Somerset, England.
Photograph © 1995 by author.

A second well, known as the White Spring, comes out of the Tor about fifty yards away from the Chalice Well. Surprisingly, it seems to have a completely different source from the Red Spring and contains very little iron and much more lime. Its volume is not as steady as the Red Spring, fluctuating between 5,000 and 70,000 gallons a day.

Clearly, a very large subterranean water system cuts through rock beneath Chalice Hill and the Tor. Celtic lore says there is a crystal castle that spirals or revolves inside the Tor, causing time distortions and sudden tunnels to open up from which no one returns sane.

The Glastonbury Chalice Well Trust in 1990 published a fact book that states: "A suggestion has been made that these well chambers were originally above ground, but in the course of time, rain and earthquakes in the 13th century brought down the soil from the hillside above and they have become silted over and completely covered.... Excavations by Mr. P. A. Raytz in 1961 (discovered) the stones on the outside of the chamber (below ground) had been chiseled to a fine finish.

"At the bottom of a trench just east of the Well, nearly twelve feet below the present surface, the stump of a yew tree was found.... It was found in a layer of blue-grey clay and clearly was *in situ* where it had formerly grown. A specimen was sent to Leeds University where it was tested for water content and found to have been a *living* tree in Roman times around A.D. 300. This suggests that the ground level at that period was approximately the same as the present bottom of the Well and that the water (in A.D. 300 and before) came out as a surface spring.

"The Well lid of wrought iron was designed by Frederick Bligh Bond, a sensitive who was the resident archaeologist of Glastonbury Abbey in the early 1900s....The two interlocking circles of the ironwork are ancient in origin and the symbolism has been found as the basis for numerous earth works and structures throughout the world. It was also used in the Sacred Geometry of Glastonbury Abbey." That sacred symbol is the Vesica Piscis.

Chalice Well and its Vesica Piscis covering in Glastonbury.
Photograph © 1995 by author.

The Vesica Piscis design on the Chalice Well is copied from a 13th century pattern and represents "the Bleeding Lance holding in balance the Visible and the Invisible Worlds interlocked with each other. It can also represent the Yin and the Yang, the meeting of the Conscious and the Unconscious and the blending of masculine and feminine energies."

Some old texts imply that Joseph of Arimathea brought Christ's chalice cup of the Last Supper with him back to Glastonbury after the crucifixion. Thus the name Chalice Well. "Chalice Hill, Chalice Well, all bespeak St. Joseph and the Holy Grail. St. Joseph and his eleven anchorites lived around the sacred well."

The Reverend C. C. Dobson, in his 1936 book *Did Our Lord Visit Britain as They Say in Cornwall and Somerset?,* even postulates that Jesus went back to Glastonbury for prayer and meditation before his earthly mission in Israel. Why would Christ choose Glastonbury?

Dobson writes, "The reason may perhaps be found in Druidism. ...The basic Druid belief was in a Trinity. It was not polytheistic. The Godhead was called Duw, the one without darkness who pervaded the universe. Three Golden Rays of Light were the emblem of Druidism, representing the three aspects or persons of the Trinity emanating from the Godhead. They were known as Beli, the Creator as regards the past, Taran, the controlling providence of the present, and Yesu, the Coming Saviour of the Future. Druidism thus *anticipated* Christianity and pointed to the coming Saviour under the very name by which Christ was called. The more Druidism is studied, the more apparent is its relationship to the revealed religion of the Mosaic Law."

When I visited Glastonbury and drank from the iron-filled water in 1995, I talked to people who live nearby. Many know and accept as true the oral and written tradition that Joseph of Arimathea went to Britain after the crucifixion with several companions and was given "twelve hides of land in the island of Ynnis Witrin," where he built a round mud and wattle church and dedicated it to Mary at Christ's request.

Ynnis Witrin means "glassy" or "crystal island." Christians did settle near the well for its plentiful water and to baptize. Today the water passes through an underground piping system and comes out above ground first at a drinking faucet and then in a double-circle "Vesica Piscis" waterfall near the main entrance.

"Vesica Piscis" waterfall near main entrance of Chalice Well.
Photograph © 1995 by author.

The Chalice Well Trust says, "The water can then be traced along Chilkwell Street to the top of Bere Lane and thence into the Abbey grounds. The Abbey in Glastonbury is now acknowledged to have been the first church built in honor of Christ and his mother, Mary, a few decades after the crucifixion. At the excellent Museum on the Abbey grounds, a plaque says construction began around 63 A.D.:

> "No one is sure when the first Christian community came to Glastonbury. Traditions tell of Joseph of Arimathea, said to be Christ's great-uncle, settling here with eleven disciples in AD 63, of St. Patrick being made abbot of a Celtic monastery here in AD 443 and of a visit by St. David. In particular, they tell of the 'Old Church' — the Vetusta Ecclesia — made of wood and mud and dedicated to the Virgin Mary by Christ Himself.
>
> "William of Malmesbury wrote in Chapter 18 of his *De Antiquitate Glastonie Ecclesie:* 'The church at Glastonbury is the oldest of all those that I know of in England.... In it are preserved the bodily remains of many saints,...and there is no part of the church that is without the ashes of the blessed.'

And Glastonbury's esteemed St. Dunstan of the ninth century A.D. wrote: "In that place at God's command the first people baptized in the catholic law discovered an ancient church built by no human skills."

Even a thorn tree grows near Mary's Chapel that does not grow naturally anywhere in the world except the Middle East. The museum calls it the Holy Thorn of Glastonbury with this small written history: "On landing at Wearyall Hill, so the story goes, Joseph thrust his staff into the ground whereupon it took root and flourished. The Holy Thorn which stands in the abbey grounds is reputedly a cutting from this tree. It flowers twice yearly, in late spring and at around Christmas time, reminding us of Christ's birth."

A species of thorn tree grows near Mary's Chapel at the Glastonbury Abbey that is native only to Israel. Photograph © 1995 by author.

Part of the legend is that after Mary died, Joseph of Arimathea took her body back to Glastonbury to be buried in a crypt that is now honored at Mary's Chapel on the abbey grounds. On the south side of St. Mary's Chapel there is an ancient stone carved with the words "Jesus Maria." The source and date are unknown.

Stone carved with words "Jesus Maria," of unknown date by unknown source on south wall of St. Mary's Chapel at the nearly 2,000-year-old Glastonbury Abbey. Photograph © 1995 by author.

Lionel Smithett Lewis, in his 1922 book *St. Joseph of Arimathea at Glastonbury,* quotes a Jesuit priest centuries ago named William Goode as asking, "Why suddenly in that wall do those two names appear? The monks evidently attached a great veneration to that stone. Did the feet of these holy beings named ever tread this spot?"

Lewis also writes, "Perhaps there is some truth in the strange tradition which still lingers, not only among the hill folk of Somerset, but of Gloucestershire, and in the West of Ireland, that St. Joseph of Arimathea came to Britain first as a metal merchant seeking tin from the Scillies and Cornwall, and lead, copper and other metals from the hills of Somerset, and that Our Lord Himself came with him as a boy."

Lewis's research confirmed that the two earliest exhibits of Roman-British lead in the British Museum came from the Mendip Hill near Glastonbury. Early in the 20th century, researchers were digging at Ostia, a Roman seaport, and found an ancient drainpipe below the chariot road. "It was bonded in some special way with tin. Professor Russell Forbes cut off a section and sent it home to England without comment for analysis. The verdict was that the metal came from the Mendip mines. ... It is also said that in the wonderful aqueduct in Jerusalem, attributed to King Solomon, the particular type of lead found in the Mendips and nowhere else was used. This was 1000 B.C."

Lewis also quotes Queen Elizabeth I from a 1651 book as stating: "Joseph of Arimathea planted Christian Religion immediately after the Passion of Christ in this Realme." Further, Lewis points out that Sir Robert Wingfield, Henry VIII's ambassador, recorded, "Immediately after the passion of Christ, Joseph of Arimathea, the noble Decurio (recognized office in the Roman Empire in the time of St. Joseph)... immediately proceeded to cultivate the Lord's Vineyard, that is to say, England, and converted the peoples to the Faith."... (Vineyard) is a curious name. Why should England be called the Lord's Vineyard? One feels bound to record that on the south side of Weary All Hill, which is the traditional landing or resting place of St. Joseph and his eleven companions, is a piece of land still called the Vineyard."

The timeline given for Joseph's move to Glastonbury varies, but A.D. 31 is the year used by Lionel Lewis. Rome invaded Britain in A.D. 43 to expand its territory and to eradicate the Druids. Lewis says that Joseph of Arimathea and ten companions traveled through Britain preaching about Christ and settled in the "Glassy Isle" assumed to be Glastonbury. Lewis writes, "After a very short time... (they) were told in a vision by the Archangel St. Gabriel to build a church in a spot revealed to them in honour of the Blessed Mother of God, the ever Virgin Mary. Obedient to the heavenly admonition, they compiled a certain chapel with circular walls of twisted twigs in the thirty-first year after Our Lord's passion."

Lewis also describes the Welsh tradition that "St. Joseph of Arimathea brought Christianity to Britain and is always called 'Ilid.' Why, has always been a puzzle. With diffidence I will suggest what may be a solution. The word *Ilu* is found at Babylon as the equivalent of the Hebrew Elohim, God. Jah-Elohim, and God, becomes *Jaum-Ilu*; Jacob, Joseph, Israel become *Yakub-Ilu, Yasub-Ilu, Sar-Ili.* On the stele of Manepta in Egypt, Israel becomes *Isra-Ilu.* Joseph or Yasub means 'he who adds.' Does the full name *Joseph-Ilu* therefore mean 'God's man who adds,' 'God's rich man,' 'God's Joseph?' In a strange land (Britain), the Yasub or Joseph might easily be dropped, and Ilid might well be a Welsh variant of *Ilu or Ili.*"

Lewis found old Roman descriptions by Diodorus Siculus, who traveled to an island in what is now Great Britain in 80 B.C., where "metal workers of Balerion or Cornwall carried their tin 'to a certain island lying off the coast of Britain called Ictis; for as the ground between is left dry at low tide they carry tin there in great abundance in their waggons. (If) Glastonbury is Ictis, the island where metal was brought for foreign export, it certainly gives support to the story that St. Joseph came to Glastonbury as a metal merchant before he came there as a missionary, and would account for friendly reception by King Arviragus (King Arthur in Glastonbury?) and the Druids, and for his choice of the spot there for his mission."

Construction of Glastonbury Abbey began around A.D. *63, perhaps inspired by Joseph of Arimathea, uncle of Christ's mother, Mary. Historically considered to be the first church in the world dedicated to Jesus Christ and Mary after the crucifixion. Photograph © 1995 by author.*

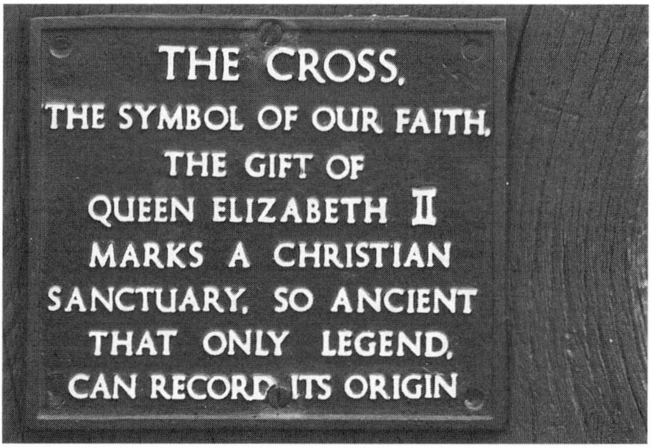

Plaque on the large cross at the Glastonbury Abbey. Photograph © 1995 by author.

By A.D. 63, construction began on the great Glastonbury Abbey, the first church in the world dedicated to Christ and Mary. Older than the

Glastonbury Abbey is the Glastonbury Tor, where Joseph of Arimathea and the young Christ walked, according to local legend, and drank from the red and white springs. In the abbey museum, there is a photograph of the sea flooding the lower land and making the Tor an island.

Early 20th-century photograph of the Glastonbury Tor surrounded by seawater after a natural cyclic flooding, displayed in the abbey museum.

This ebb and flow of seawater inland has happened repeatedly for as long as anyone can remember, and explains why the legend of King Arthur and his Round Table of Knights lived, fought and died in the "mists of Avalon" — another name for Glastonbury when water filled the surrounding lowlands.

Britain's National Trust describes the Glastonbury Tor this way: "'Tor' is a west country word meaning a hill. Glastonbury Tor is a natural formation composed of layers of clay and blue limestone, capped by a mass of hard, erosion-resistant sandstone. Its top is 518 feet above sea level. The tower is all that remains of a medieval church of St. Michael built by the Abbey to replace a previous church which fell in an earthquake in the 13th Century. Excavation of the summit has revealed traces of much earlier buildings, dating from the 5th or 6th Century A.D.

"The Tor is and has been to many people a place of magic, the focus of legend and superstition. There is a current theory that the terraces formed a sort of three-dimensional maze. One local story says that there is a hollow space inside; another, perhaps more ancient idea is that the hill has a secret entrance to the Underworld.

"The Tor was the scene of the hanging, drawing and quartering of Richard Whiting, the last Abbot of Glastonbury, when Henry VIII dissolved the Abbey in 1539."

The Tor and St. Michael's Tower rising above Glastonbury, Somerset, England. Photograph © 1995 by author.

John Michell, an authority on ancient science, religion and the symbolism of sacred landscapes, wrote in his 1990 book, *New Light on the Ancient Mystery of Glastonbury:* "Behind all the legends, prophecies and revelations at Glastonbury can be discerned one single theme: that the will of God will finally prevail and humanity will rediscover its natural condition within an earthly paradise.

"The puzzling, unfamiliar yet wonderfully exciting conclusion which is forced upon us by examination of the Glastonbury mystery is that we are being asked to contemplate an alternative reality, a reality quite unacknowledged by modern habits of thought, the reality of a heavenly order on earth. That is the ideal which appeals to and links all humanity. It may now sound unreal, but it was not considered so in many different ages in the past, and it seems beyond doubt to be the only practical ideal and image for the present circumstances....The secret of Glastonbury is that we are invited to enter an earthly paradise."

Celtic cross surrounding six-petalled flower in stone stands in front of a downtown Glastonbury church. A six-petalled flower is also on the ancient Glastonbury seal shown below in the Abbey museum.

Since the summer of 1998, there have been at least fifteen crop formations in the Somerset County area, and four have been variations on the Godhead Trinity symbol, whether by human or other design.

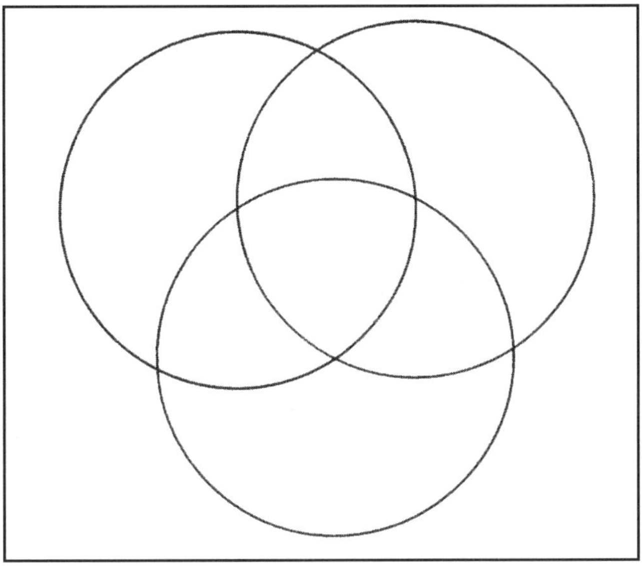

Trinity symbols were discovered in North Somerset at Marksbury near Bristol on June 19, 1999; Penny Quick Hill near Bath on June 20, 1999; and Corston near Bristol on July 3, 1999. Diagram by the author.

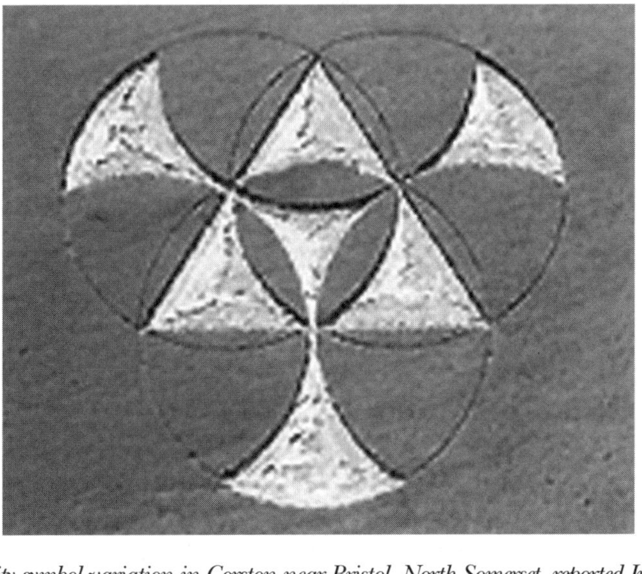

Trinity symbol variation in Corston near Bristol, North Somerset, reported July 4, 1999, in wheat, about 100 feet in diameter. Photograph © 1999 by Peter R. Sorensen.

6
Ghostly Images

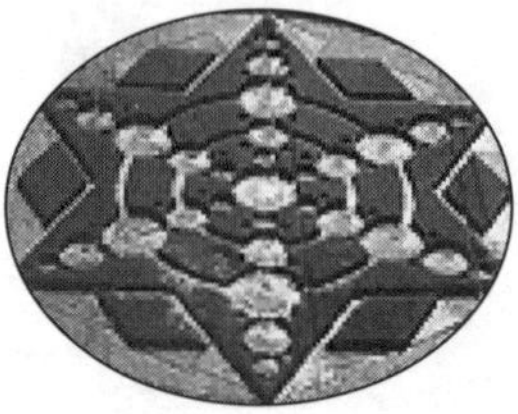

When we were on the hill above Roundway, Charles and Frances Mallett had shown me two photographs taken in the East Field at Alton Barnes below Adam's Grave after the two biggest and perhaps most spectacular formations of 1999 appeared on June 12 in barley. Charles had been there all that night of June 11–12 only about 150 yards from where one of the two formations occurred. Charles told me, "The night had been a dead-still, perfectly quiet night where you can hear rabbits running all through the barley. Later in the night, a mist comes down over this field, a localized mist. There are people on Knap Hill with searchlights scanning this field for quite a bit of the night while I'm awake. Those people were actively looking for hoaxers in the East Field. I went to sleep very close to the edge of the field.

"I wake up in the morning around 5 AM, sit up and rub my eyes, and there is a crop formation right in front of me! I stand up and there is another one to my left, the massive one. I'm the first person in these for sure, no doubt. Obviously, I'm really excited, pack myself up and jump over the wire. And I'm in there quickly.

"As I approached these two crop circles, the 'Serpent' was in front of me so I went into it first. I'm experienced in looking at crop circles. I look for signs of other people being there. As I walked to this, it's in very young barley and there's a heavy dew on the crop. My legs are soaked, my boots are caked with mud by the time I get to the crop formation. I'm leaving very muddy footprints all over the place. I'm examining the stalks and lay of the crop for mud prints. Mine are the only dirty boot prints there."

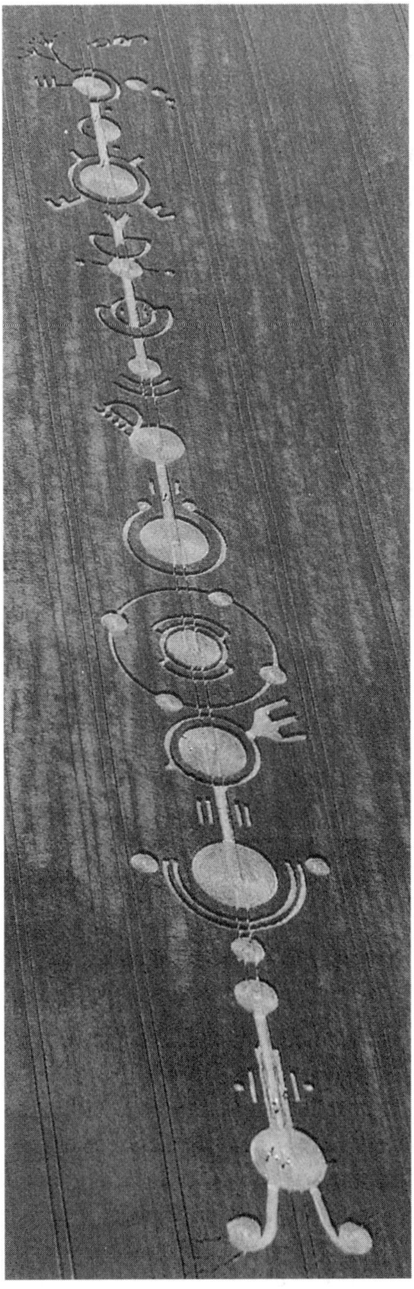

"Sentence" on left measured 1,020 feet long with signature at far end; "Serpent" on right was 500 feet long. Both discovered at 5 AM on June 12, 1999, by Charles Mallett. Aerial photographs © 1999 by Lucy Pringle.

Charles said, "I'm convinced absolutely 100 percent that no one else had been in those crop formations during the night. The dew comes down early. If anyone else had been in there, they would have left dirty footprints everywhere. There were standing rows of barley with absolutely no damage. It was pristine, precise, clean-cut. Five hours of darkness and those two arrive. Impossible by hand, I would say.

"The pictogram was amazing, over a thousand feet long. The other one, snake or serpent — or whatever you want to call it — 500 feet long. I'm even in the right tramline! The serpent is right in front of me about 150 meters."

"Charles, anybody sleeping only 150 meters or about 500 feet from where two massive crop formations were made should have heard noise and woken up, shouldn't they?"

"I don't know if I would have woken up or not, but whilst I was actually observing the area awake, you can hear rabbits and animals running through the crop. As a comparison, we slept in the long one — the pictogram next to the snake — the following night. It was another still, quiet night. We could hear people at least a half mile away chatting at normal levels. If there had been a team of people in the East Field, they were so coordinated they didn't even have to speak to each other."

"And, Charles, you said you slept for only about five hours. In the hoaxes, it has taken a team of people hours to do a simple formation."

As the three of us looked at the photographs of the East Field formations, Frances pointed out that the small signature at the end of the Sentence was similar to the signature in the Cherhill formation. It had a curved J coming out of a central circle, but instead of an F, the signature appeared to be topped with a knobbed ring.

The two massive and complex formations in the East Field on June 12, 1999, appeared as if the phenomenon had raised the ante in whatever game was being played. Whatever energies were required to make them seemed to linger long enough afterward to interfere with nearly everyone's cameras and battery-powered equipment.

Frances told me, "People's cameras were going down right and left. We have a friend from Surrey who brought his camcorder. He had two brand new battery packs, sixteen hours' worth of use. Each one of them went down in three minutes, sucked of all energy.

"We decided to sleep that night in the small signature at the bottom of the long formation. Our dog, who has been in all the formations since he was a puppy, really did not like this one. He was cowering down as if he either had something oppressive above him or he was in a tunnel of some kind. He kept seeing things we couldn't see. Kept focusing on things, growling, barking and putting his face really close to us. He didn't want to leave us. He was sick all night."

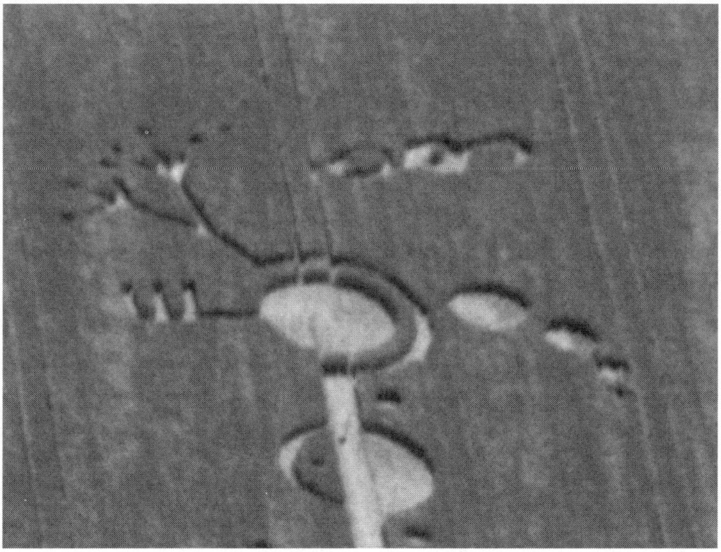

Photo enlargement of the signature detached from the long formation, which resembles Cherhill 1999 and Dadford 1998, but has a different ringed top. Aerial photograph © 2000 by Lucy Pringle.

Frances paged through her binder and pulled out a photograph of the long Sentence, which had a strange image of portholes or a pea pod superimposed over the field at one end of the long pictogram.

Frances explained, "Francine Blake took this photo and it has this very strange anomaly taken in the daylight. What on earth is it?"

Francine Blake started the Wiltshire Crop Circle Study Group (WCCSG) in 1995 and edits *The Spiral* newsletter. Charles added, "It looks as if it's something that is just tuning in almost — the front of it has got in and the back end is just about phasing in. Quite weird."

June 14, 1999, aerial photograph of East Field "Sentence" with "porthole" anomaly in foreground. Photograph © 1999 by Francine Blake.

Francine Blake sent me the above aerial photograph along with another she took on the ground in the Sentence formation on Sunday, June 13, in the early morning. The second photo has a similar anomaly. With the photographs, she sent the following report:

> I arrived at 6:45 AM on Sunday morning (June 13, 1999) to find quite a few people already there. I decided to go into the long pictogram first and the couple [Charles and Frances Mallett] who were walking in front of me, but in a parallel tractor line, entered the formation only to turn back immediately. This puzzled me and I shouted, asking them why they had not stayed in. They replied that their dog, which I could not see, had become too distressed to enter the formation.
>
> I sensed a strong atmosphere on entering the pictogram and started to feel somewhat queasy myself. I began to photograph the long formation and at its far end took a picture of the attendant serpent a little distance away. When I received my photographs a few days later, I was astonished to see light anomalies on them. The negative of the film taken on Sunday showed a

progressive darkening throughout, culminating with a startling light shape in the last frame. I could not see anything of this with my eyes inside the formations."

Anomalous photograph taken early Sunday morning, June 13, 1999, as Francine Blake walked in the barley to photograph the "Sentence" and "Serpent" in the East Field, Wiltshire. Photograph © 1999 by Francine Blake.

On Monday, I got airborne at about 2:30 PM to take aerial shots of the two formations using the same camera, but with a new film. As on the previous day's film, the shots get progressively darker showing curious light patterns around the formations, again culminating in a very strange looking shape in the last frame which is very similar to the one photographed on Sunday in the same place within the field, but taken from a different angle (on the ground).

I took the negatives back to the lab where they had been processed. The technician told me they had been processed with a batch of other films which had come out perfectly all right and that he could not come up with any technical reasons for the presence of these anomalies. He said that if the films had stuck together, or to the tray, it would have prevented the chemicals getting to those areas and the effect would have been the opposite of what it is at present. In other words, the shapes on the negative would have been left transparent instead of being black as they actually are. This would allow the light to stream through during development making dark shapes on the photographs instead of light ones. He cannot understand how the chemicals could have reacted so differently on my films from the other films in the batch which came out perfectly. He can only deduce that something from outside *did* make an impression on the films or that somehow some force affected the chemicals in the processing tank to precisely enhance certain localized areas on my films only, a supposition that gave him goose pimples!

I use a Mamiya 645, a middle format semi-professional camera. I use 220mm films (30 frames) ISO100 which I buy from a shop that supplies professional photographers. These were new Fuji films bought just a few days before the appearance of the double formation. I keep my films refrigerated (cold, not frozen) and they are sealed and in their original boxes. No one else has access to my films and I loaded the camera myself just before flying. I have used the same camera, the same type of film and the same laboratory for the past two years. I take the aerial shots from a small helicopter especially designed for aerial photography. I have taken hundreds of photographs in the same conditions and never had results like this. The days the photographs were taken were dry and sunny, eliminating the possibility of moisture on the lens.

A week later, at 6 AM on June 19, crop circle researcher Frank Laumen from Leverkusen, Germany, and friends drove to Wiltshire and the East Field to see the remarkable long Sentence and Serpent formations. He said it had been clear and sunny until they reached "the border of Wiltshire. After that, we saw little because it was so foggy. The fog disappeared very slowly, but at 8 AM we saw the large East Field formation. I was absolutely impressed! In the last years since 1997, the first year that I saw crop formations, I had hoped to see such a formation. It was so special because it was a compilation of designs over the past ten years.

"I went into the formation at 8:30 AM. It was overcast and every minute we thought it would start to rain. I took a lot of pictures, but I saw no anomalies. Between 9 and 10 AM, I made a lot of pole shots with my still camera and videotape camera. We left the formation at noon.

"When I was back in Germany, I looked at my slides and I discovered the strange orange light in one of my pole shots in the East Field." Poles are used to hoist cameras up to get a low aerial view of formations without the expense of hiring an airplane or helicopter. The anomalous bright light is above the far end of the formation in the enlargement below made from Frank's 35mm color frame. The pale spot below is unidentied and might be a film defect.

Enhanced enlargement of Frank Laumen's photograph in East Field near Alton Barnes, Wiltshire, on a grey overcast morning between 9 and 10 AM on June 19, 1999. Anomalous light and blue spot at the end of the 1,020-foot-long "Sentence" formation. Pale spot below unidentified. Photograph © 1999 by Frank Laumen.

In addition to the unexplained spheres on her Stonehenge slides, Janet Ossebaard from Gieten in the Netherlands found other ghostly images on 35mm slides she took at two major 1999 formations. We were talking one afternoon in the parking lot of the Henge Book Shop in Avebury. Janet told me about a pole shot she had made in Devil's Den. She said there was a triangular section that was lighter than the rest of the photograph.

Unexplained triangular section that is lighter than the rest of the Devil's Den photograph taken on July 20, 1999, a day after the formation's discovery. Photograph © 1999 by Janet Ossebaard.

"How long after Devil's Den was discovered?" I asked.

"It was the second day afterward. I took many photos and in one of them is this column or odd patch of light."

"Was it the only frame that was anomalous?"

"In my Devil's Den photos, yes. But I got another strange photo when I flew over the Hackpen Hill formation in 1999 and took a lot of aerial shots. One has a grey streak or beam. In all the other shots, there is nothing. The shadowy beam comes from out of the formation and straight into the clouds. It's when we flew away again to another formation and I turned back to make the last shot. You see the formation on the ground in the landscape. The beam is thin, but very long."

Unexplained shadowlike beam extending upward into sky from July 1999 Hackpen Hill formation. Aerial photograph © 1999 by Janet Ossebaard.

On July 27, 1999, Bert and Janet had joined Ed and Kris Sherwood and others at Oliver's Castle near Roundway for a meditation and night-watch. Janet said, "After about thirty minutes, we ended the meditation and started making photos and slides. It was a chilly, dry summer night. There was an almost full moon. I took slides with my Pentax MG camera using a 200 ASA film."

Janet did not see a light at the time, but when her film was developed and printed, a bright light was visible on the left side of the tree. The full moon was on the far right of the tree.

Janet at first reasoned, "It was probably the camera flash of somebody standing on my left side taking a picture at the same moment I did. But back home in the Netherlands when I projected my slides and looked at them carefully, I noticed there was no silhouette of a person holding a camera causing that flash. There should have been a silhouette. But the light is just hovering in the air. It was not attached to anything or anybody."

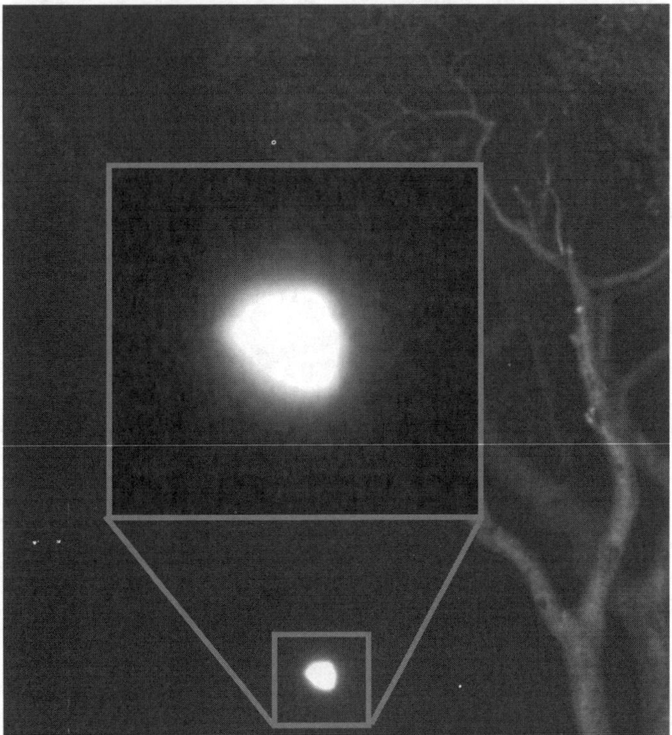

Oliver's Castle, Wiltshire, on July 27, 1999. Janet Ossebaard aimed her camera toward the tree. No light was visible, but it appeared on print after film was developed in the Netherlands. Photograph with enhanced blow-up © 1999 by Janet Ossebaard.

"Do you have any firsthand knowledge of lights or spheres showing up in a field right before or after a formation?"

"No, not really. But something strange happened in July 1995. Bert and I were doing a nightwatch on top of Knap Hill and around 1 AM, we both heard a very strange electrostatic sort of sound, the same sort of sound you hear when you drive underneath power lines. This electrostatic sound was very strong and definitely there, and it was moving. At first it started behind us, like someone walking around you in a circle and observing you. But you can't see it, you just hear it. And this sound moved and disappeared and came back again in a different spot as if moving all around and observing us. I had a very strong sense of being observed.

"It was like my mind was being manipulated because I was there for the nightwatch. I was there with my camera, with my recorder. All I had to do was press a button to record the sound. And I didn't do it because it didn't come to my mind. And that's really weird! We're trained to do this. It's our purpose. It's why we're there. We have all the equipment ready, yet we did nothing.

"And this sound was just moving around us and observing us and I felt very happy. Very safe and very secure. And all I could think of was this one line I was sending out towards the sound — I know it sounds crazy, but I'll tell you the way it went anyway: All I thought was, 'Ah, welcome! I'm so glad we finally meet.' And that was all I could send out. There was a very strong sensation of contact. This went on for minutes. Then, it disappeared and eventually we fell asleep.

"The next morning, around 5 AM, it was dawning. We got up and we didn't talk about it, I guess because it was just too strange. We were walking down Knap Hill. At the bottom there was a fresh grass circle. And we were absolutely certain it was fresh because the night before, we'd checked the grass there. We found two little grass circles about one meter in diameter. We took photographs and checked the rest of the grass in the field to see if there was anything else. And there was nothing. Just those two small ones.

"So now, after the night of the electrostatic sound, there was a third grass circle. We checked it with our dowsing rods and it reacted very strongly. So something happened there that night. Except we didn't see anything. We didn't see balls of light. We did not see UFOs or whatever. We saw nothing."

"Did you try taking any photos that night?"

"No, that was the most stupid thing!"

July 1995, third in a series of small fresh circles that appeared at the base of Knap Hill after a nightwatch on which Janet Ossebaard and Bert Janssen heard strong electrostatic sound moving around them. Two other circles were found the night before, this third in the early morning. Photograph © 1995 by Janet Ossebaard.

There are also variations on light mysteries, such as beams that have been seen in Wiltshire.

For example, one evening near Alton Barnes in the fall of 1996, photographer Peter Sorensen and a friend who wishes to remain anonymous saw two strange beams emitted from a stubble field after harvest that moved back and forth. The beams were translucent, curved, and had a flexible, rubbery motion. For this book, Peter sent me a computer-generated image of what he remembered with this note: "The sighting occurred after the 1996 circle season was over. It was already well into the evening when we went to a secret location in the vicinity of Alton Barnes where my friend had seen anomalous lights on previous occasions.

"It was a wonderful Wiltshire night with thousands of stars in the crystal clear sky, but no moon. There was a heavy, low lying mist about a foot deep that covered the stubble of a large, recently cut field. The mist glowed with a strange intensity as if there was a full moon. I remarked that it was hard to believe that the stars alone could illuminate the mist so brightly. My friend said he thought it was energy in the field itself that caused the glow. He had seen it there before.

"We stood at the edge of the field enjoying the stars as people do who live in cities and have no idea how beautiful the Milky Way is.

"It must have been nearly two hours later with nothing unusual seen. We were getting tired and about to leave when it happened. Two beams of light shot up from the middle of the field looking very much like wartime searchlights hunting for planes. The beams went up into the sky hundreds, if not thousands, of feet waving lazily to and fro. They looked like something out of a WWII movie, except they were flexible, bending like the beams were made of translucent rubber!

The curved, "rubbery" moving beams Peter R. Sorensen saw twice in a plowed field near Alton Barnes in the fall of 1996. Computer-generated image © 1999 by Peter R. Sorensen.

"The curved beams persisted scanning around with the distant end of their cones lagging behind the brighter bases for about seven seconds, long enough for us to be sure of what we saw.

"My friend exclaimed, 'What is this? Some kind of ghost?!'"

"After we recovered from our amazement, we hurried to where the lights had emerged from the field, but there was nothing to be found.

"We waited around for another hour or so and our patience was rewarded with another, shorter display again at the moment we decided to leave. It was as if the lights heard us and said, 'Farewell!' Of course, that made us stick around even later, but we saw nothing more.

"There had been no sound and no static electric tingle or other unusual phenomena during the entire period we were there. What we saw was 'impossible.' Normal light beams are caused by photons leaving a light source at 186,000 miles a second, illuminating dust particles in their path. So, the sides of light beams are perfectly straight. Physics as we know it would require a tremendous gravitational field to bend a beam of light. Perhaps the beams were actually some form of glowing plasma?"

7
Clouds of Light

On August 3–4, 1994, Werner Anderhub from Bern, Switzerland, and a friend decided to visit a crop formation at midnight several days after it was discovered on July 23. It was called the "Galaxy" at West Stowell, a little east of Alton Barnes.

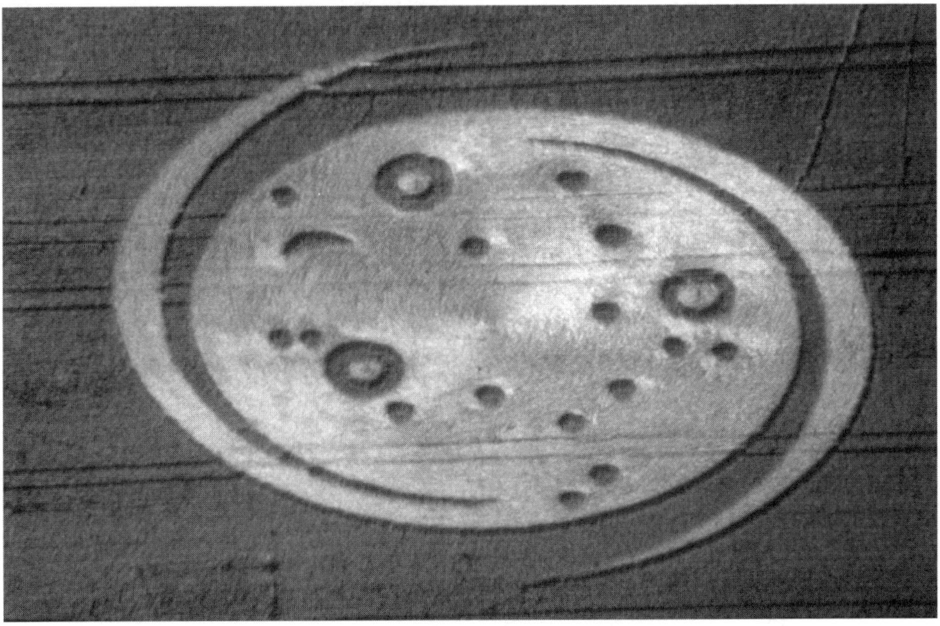

West Stowell, Wiltshire, "Galaxy" discovered July 23, 1994, in wheat. Photograph © 1994 by Chad Deetken.

As the two men entered at the southeastern corner of the field about 150 meters from the Galaxy, a very large and bright light seemed to come from the ground inside the crop formation and rise into the sky. Werner told Bert Janssen and Janet Ossebaard in their 1999 documentary *Crop Circles, The Research,* "It was so bright, it lit up the hills in the background. It was bluish-white in color and about as big as the formation, fifty to sixty meters. The bright light formed some sort of cloud and it changed shape continuously as it hovered over the formation. After a couple of seconds, it rose at slow speed and disappeared into the darkness. My friend and I were totally flabbergasted!

"As we stood in the dark night, a second light emerged from the formation, repeating exactly the same action and movement as the first one. And after about a minute, there was a third light. And then a fourth light appeared.

Inside Galaxy formation at West Stowell, Wiltshire, after its discovery on July 23, 1994. Photograph © 1994 by Chad Deetken.

"This fourth one did not just form a cloud. It formed a perfect rectangular shape, like a big shoe box hovering only a few meters above the wheat and measuring five by twenty by sixty meters. It was transparent like a jelly fish. And it didn't just hang there. It started to float towards us.

"At that point, my friend began to get really scared and wanted to run away, but I asked him not to go. At that instant, the light stopped. It floated back towards the formation where it transformed into another cloud and moved slowly up into the dark sky.

Drawing with note: "The five sequences of the light encounters began all in the same way as a fluid light came out of the Galaxie, forming a cloud above the field, stayed there for several seconds, moved slowly up into the sky, and then disintegrated into nothing visible."
Drawing by Werner Anderhub for author © 2000

"At that point, two other friends arrived and a *fifth* light emerged from the formation, forming another bright cloud that hovered for a couple of seconds after which it moved upward and disappeared into the sky.

"We had witnessed something that could not be easily explained. What struck us most was the strange rectangular box, how it approached us and then how it left when my friend got really scared. It was as if the light were intelligent, or guided by an intelligence."

"The 'cloud' disintegrated by moving up into the sky and lighted up the Vale of Pewsey. We could easily see the Golden Ball Hill with its bushes and sheep. Please note this was not *a UFO. I have seen those. This was a totally different event." Drawing by Werner Anderhub for author © 2000.*

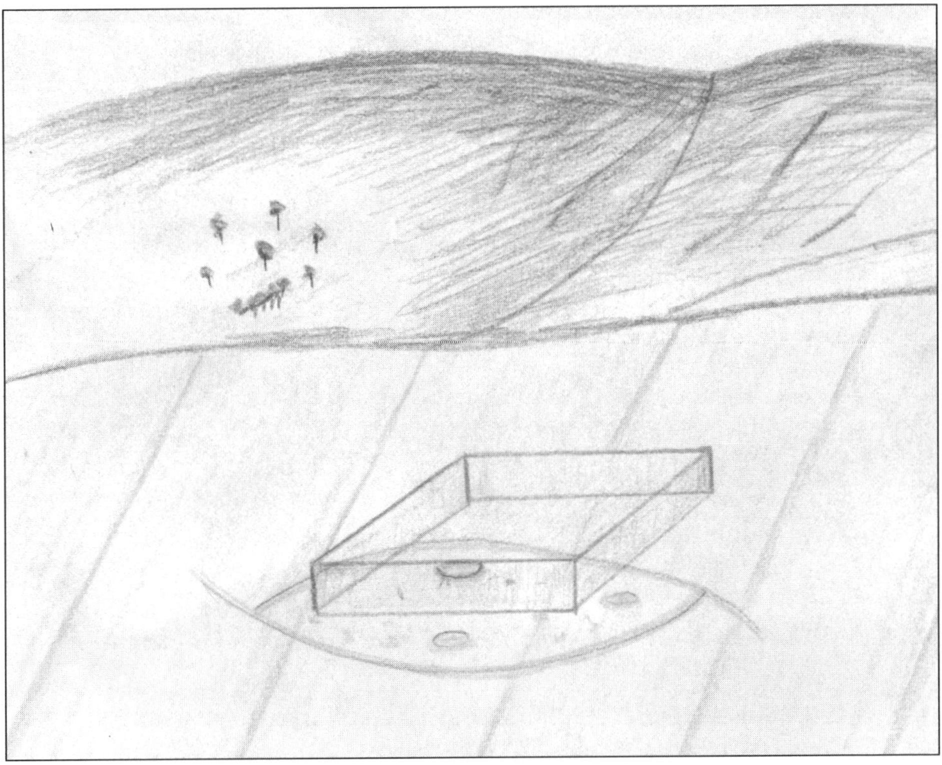

"After the 'light cloud' was formed above the Galaxie formation, in an instant the light cloud became the form of a 'shoe box' hanging over the field for awhile. Then it moved slowly in our direction."
Drawing by Werner Anderhub for author © 2000.

A week earlier, on July 16, 1994, an elegant large ring trailed by eighteen circles and three grapeshot were found at Wilsford south of Alton Barnes. That day I was in Baltimore, Maryland, at the Space Telescope Institute interviewing the late lunar geologist Eugene Shoemaker and MIT astronomer Heidi Lammer about the impact of twenty-one comet fragments one after another into the giant gaseous body of Jupiter. I was with the scientists on July 16 when the first collision was photographed by the space telescope. Champagne was opened and passed around in the excitement of actually witnessing an astronomical event so rare it might have happened only a few times in the entire life of the solar system.

Could that large ring and twenty-one trailing circles at Wilsford, also arriving on July 16, 1994, have been another intelligence's way of saying it knows what is happening in the cosmos beyond earth?

Wilsford, Wiltshire, in wheat discovered July 16, 1994.
Aerial photograph © by Chad Deetken.

Two visitors to the Wilsford formation were Foeke Kootje and Conny de Bruyn who founded an organization called UFOCUS in the Netherlands to study crop formations and their possible connection with unidentified flying objects and lights. On July 25, 1994, at about 1:15 PM, Foeke and Conny arrived at the Wilsford formation nine days after its original discovery. Wilsford was the first formation they had seen up close.

Foeke told me, "We went on the inside of the formation. I filmed inside as well. Later on, when we came out, we decided to film the formation from the road on a tripod. So I positioned my camcorder on the tripod, zoomed to close up on the formation and pushed the record button. While I was filming, I didn't look all the time through the viewfinder because it was a good steady shot without shaking. I recorded maybe twenty-five seconds and shut the camera off.

"Later on after the holidays and we came home, we looked at the footage and discovered this strange light jumping around in the formation. I had the impression it had been there for much longer than I recorded on camera. We didn't know this object was over there while we were filming. That disappointed us because we would have recorded much longer than just fifteen seconds. But we were very lucky, I think, to capture this light on video. This was proof for me there is more going on in the crop circle phenomenon than just manmade formations."

It's important to underscore that neither Foeke nor Conny saw any light activity in the Wilsford formation while they videotaped.

Foeke was truly surprised when he looked at what had been photographed. "In the footage, you can see a very bright light moving inside the formation in the left corner below the big ring in the standing corn. First, the light looks like it just appears in the air or just jumps out of the circle. It's difficult to know for certain. Later on, it moves from left to right and back from right to the left.

"Also, when we saw it for the first time on our original tape, you could actually see more strange little dots or balls flying very fast around in the formation. In the top right corner of the big ring, the tape shows a strange glow and in the whole formation you see fast flying dots of light, which you can clearly see if you play the tape frame by frame. These strange dots have been seen by several people. We have checked other footage we have made in different formations, but we could not find the strange flying dots on any other tape. We are sure there were no other people in the formation while we filmed this strange happening."

Foeke Kootje and Conny de Bruyn videotaping inside a Wiltshire crop formation. Photograph © 1999 by John Holman.

"And what color were the flying dots?"

"More or less white."

"How big?"

"Very small. More like a seed blowing in the wind."

"If the formation had been there for nine days, why would the lights be there?"

"That's a good question. While I was filming the formation from the road, I was talking about the formation — whether it was real or not. From the inside, it was already quite old. I talked about that while I was filming, in Dutch. I was talking about it to myself whether it was real or not."

"So the phenomenon might have answered you and your question?"

"Maybe, yes. When I see the 1994 tape and I hear my voice talking about whether it's real or not, it's funny that you suddenly see those lights jumping around. It might have been a reaction to what I said."

Video Plate 1: Wilsford formation without light around 1:15 PM on July 25, 1994. Video plate © 1994 Foeke Kootje and Conny de Bruyn.

The following six frames were transferred from a third-generation NTSC video dub made from one of Foeke's PAL dubs of his PAL master. The transfer was from the videotape to a CD-ROM for frame storage. They are not six sequential frames, but six frames representative of the light's movement back and forth close to the standing wheat inside the large ring at the top of the Wilsford formation. The light appeared and disappeared twice over 2.5 seconds, first moving down toward the camcorder, disappearing, then reappearing and moving down and then up again.

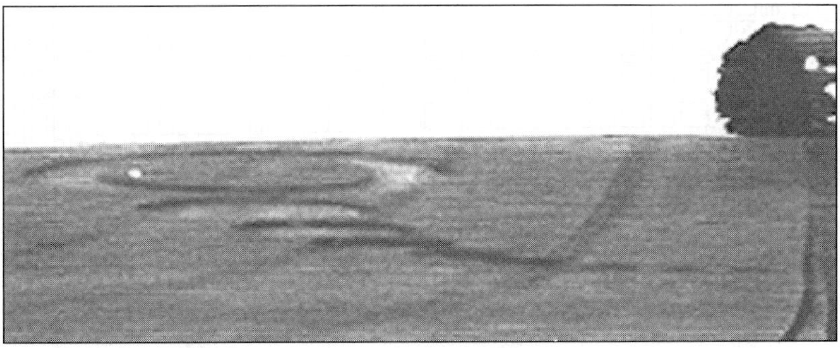

Video Plate 2: Wilsford, light first coming out of standing wheat inside large upper ring. Video plates © 1994 by Foeke Kootje and Conny de Bruyn.

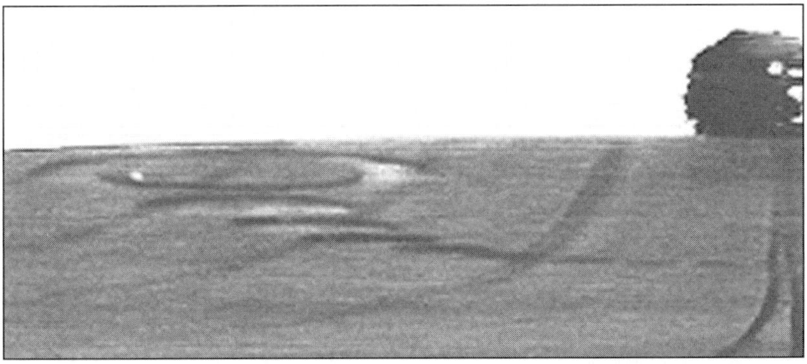

Video Plate 3: Wilsford formation with light moving toward camera.

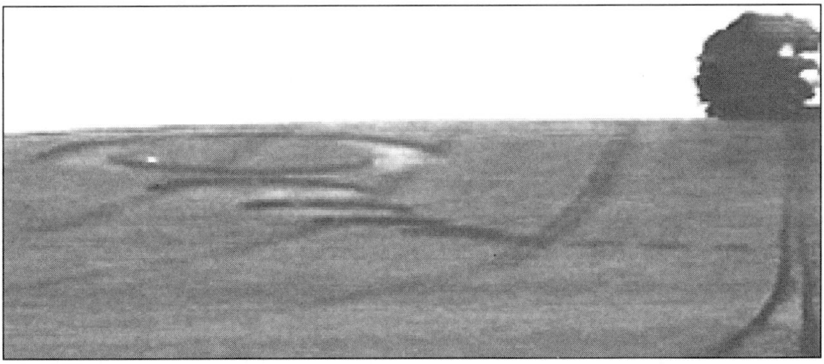

Video Plate 4: Wilsford formation with light moving toward camera.

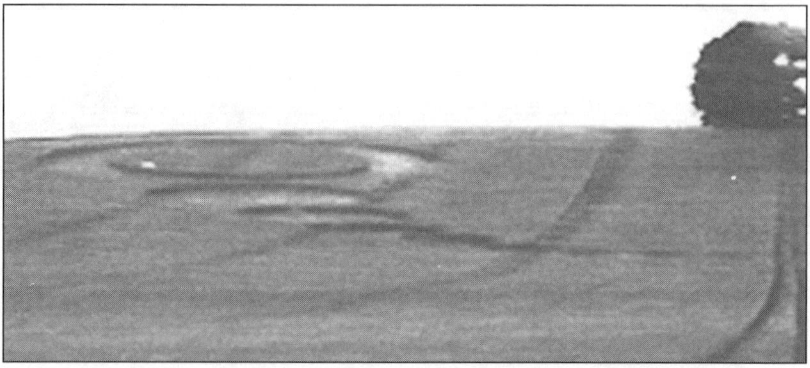

Video Plate 5: Wilsford formation after light first disappeared, then reappeared, moving toward camera again.

Video Plate 6: Wilsford formation as light reversed direction and moved away from the camera.

Video Plate 7: Wilsford formation as light moved away from camera and disappeared for a second time, July 25, 1994.

Foeke told me that while they were visiting England in 1993, they decided to climb to the top of Adam's Grave next to the East Field in Alton Barnes for a nightwatch on August 6 at about 1 AM.

He said they were "looking towards Woodborough Hill, over the East Field, when we suddenly saw a strange light flying low above the ground coming towards us. And we think it came from Tawsmead Copse near Woodborough Hill. We started to film this object on our camcorder."

"What color was it?"

"I only saw it through my viewfinder and didn't see colors. But Conny saw it better than me. She said it was one red light and two white lights."

"And it's coming toward you?"

"Yeah, towards us, and it flew in half a circle in front of us."

"How close to you?"

"I think a mile away. This object was on the forest end of the East Field, so it's above the road which runs there about a mile away."

"You were on top of Adam's Grave watching and a mile away over the tree end of the East Field; the lights did a half circle?"

"Yeah, that's right. I saw the light coming and I started the tape because I wanted to catch it on camera. If it's not interesting, I can always erase it, you know? I saw a light coming towards us and thought, 'Let's film this. You never know what it is.'

"It could have been a helicopter or airplane, those kinds of things. But when we saw it further on — we could not hear any sound. That's the strangest thing, because there normally has to be a sound if it's a helicopter or airplane."

"Foeke, on the videotape it seems to be many moving lights. What happened?"

"First, when we saw the light, it started with only one in the distance. And when it came more near to us and it flew in a half circle in front of us, the lights changed in some sort of way. I don't know. Then it was two lights and then suddenly a third in a triangular shape. You can actually see it on video. But it has to do with the direction it flies and the angle you are looking at it, I think. And when it's flying more in a half circle in front of us and goes more in the direction of Woodborough Hill from us, then the lights are doing very strange things which I can't explain. I don't know. I've never seen those kinds of lights."

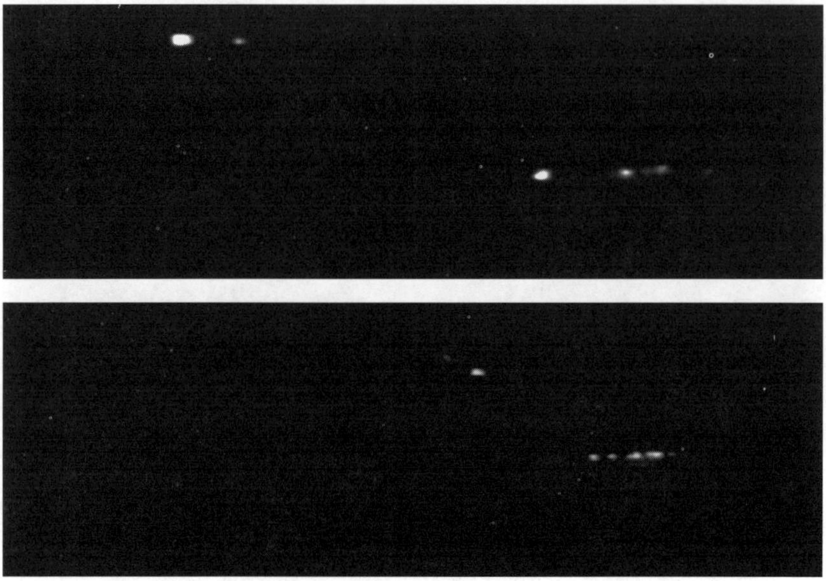

Two video frames enhanced from Foeke Kootje's tape recorded at 1 AM on August 6, 1993, when camera was aimed at Tawsmead Copse below Woodborough Hill. One red light and two white lights were visible, but computer enhancement revealed many more lights in odd pulsing and jumping patterns throughout all of Kootje's video frames. Video frames © 1999 by Foeke Kootje.

Foeke told me he guessed the various lights were on one big object, but wasn't certain. "If you see the footage," he said, "it looks like the lights are circling."

"If those were lights on one object, how big do you think it was?"

"Good question. How big is it? That's not so easy to tell. Big, I think."

"And Tawsmead Copse is where the light seemed to come from originally?"

"Somewhere there, yes. Later on, one light disappears and comes back later. What it actually is, we can't say. Maybe it could be some sort of secret airplane? We don't know for sure. But we couldn't hear any sounds. When there's any wind you can hear it in the microphone on my camera, but there was no wind. No sound at all. Nothing."

"And how did the lights disappear?"

"Well, I think later it flew behind something because suddenly you don't see anything anymore."

"How long did you and Conny stay on Adam's Grave?"

"I think for a half hour after that. I think we left at half past one."

"And you didn't see any more lights after that?"
"No."
"Were any new formations found there the next day?"
"No, there weren't."

Bert Janssen and Janet Ossebaard also videotaped unidentified lights in the Tawsmead Copse and Woodborough Hill area on August 5, 1999. They were on top of Woodborough with Ed and Kris Sherwood, Peter Sorensen, Ron Russell and several others watching the sun go down. The evening was particularly warm and beautiful.

Videographer Bert Janssen and writer-photographer Janet Ossebaard videotaping crop formations in Wiltshire, England. Photograph © 1999 by Frank Laumen.

That morning the five-pyramids formation had been discovered at the West Kennett Long Barrow that Andreas Mueller, his colleagues and I went to after the cell phone call to the Barge. I had planned to go back up on Woodborough that night as well after seeing the strange blue sparkler light and sky blue sphere that popped into view, traveled a short distance and popped out again like someone had turned the light off.

But I was diverted by the challenge of assembling a report with new photographs for my website, www.earthfiles.com, and ended up writing instead of nightwatching.

Later, at the Golden Swan in Wilcot, Ed Sherwood told me what happened to him that August 5 evening atop Woodborough Hill with the others.

"As I got closer, I noticed there was a group of people up there. And I noticed they all seemed to be looking in one direction, which was the fading sunset. And also, I started to see nonphysical balls of light materialize over them at close range, which means to me from my experience that they are all meditating. It appeared the group was meditating on the fading sunset. They were all giving their attention to the sun at the same degree and that's a form of meditation.

"So I didn't want to disturb what they were doing and I went to the far end of the hill, tried to focus on the grid of the land, and immediately sensed two locations — one at Martinsell Hill and another hill at the end of the hill chain headed south. The last hill was where I immediately began to sense that nonphysical energy was going to materialize. This is what I felt before I saw anything. Then I started to see nonphysical balls clustering at both locations.

"I went over to the group and told everyone to look at Martinsell Hill and the most southern hill. Then two balls of light appeared and disappeared over Martinsell and then at the other hill. Over a two-hour period between 9 and 11 PM on August 5, many plasmas materialized between those locations. They all had the same appearance — gold light, which is something I rarely see as nonphysical or physical energy. And after about fifteen minutes, many helicopters began to arrive from various locations, mostly from the south.

"I saw through binoculars five helicopters surrounding one object twice, maybe three times. We saw in that two-hour period maybe twenty-five or thirty objects. Bert videotaped at least three of those lights."

August 5, 1999, two video frames enhanced from Bert Janssen's tape while several unidentified gold-colored lights were seen by Janet Ossebaard, Ed and Kris Sherwood and several others on top of Woodborough Hill. Computer enhancement and frame-by-frame analysis revealed many more lights in odd appearing and disappearing patterns throughout Janssen's half hour of video frames. Video frames © 1999 by Bert Janssen.

In July 1996, Bert and Janet had seen an amber light over Tawsmead Copse for several minutes.

Bert said, "We were sitting in the car at the foot of Golden Ball Hill just beside Knap Hill. You can drive up there, if you know the roads. We were overlooking the West Woods field. Just ahead of us was Tawsmead Copse and to the right is Woodborough Hill. We had been there for about two hours. It was half past 1 AM and I got really tired and said let's go to bed. Then just in that moment that I wanted to start the car, I saw straight ahead of me at the foot of Tawsmead Copse, just at the edge of the woods, an orange light. I told Janet, 'Look what's there!'

"I took my binoculars and looked and it's very difficult to determine the size, but I could see the light very clear with binoculars and you could see it with your naked eye. After about five seconds, then it got less. It was still there, but you couldn't see it anymore with the naked eye. You could just see it with binoculars. The light was not a real steady light. It looked much more like a fire. The light was alive like a fire is alive."

"Flickering?"

"No, not flickering. It was getting more and less intense, like a heartbeat."

"The light had a pulse to it?"

"A steady, irregular pulse. It was not regular, but more like an open fire, you look at the flames."

"Was it the color of fire?"

"Yes, it reminded me very much of fire. Only it was shaped like a round torchlight, but much bigger. I think it would be the size of a soccer ball, that big. But it's so difficult to say how big a thing is when it's at such a distance in the dark. I'd compare the size to somebody's head and that's about the size of a soccer ball, I think."

"Could you tell how far away it was?"

"Well, I think it would easily be 600 or 700 meters. You have the slopes of Woodborough Hill in the background. This light was ahead of the slopes, so I know exactly the distance. It cannot be any further away than Woodborough Hill. And after ten seconds or so, it got brighter and you could see it with the naked eye again. That took about two, three or four seconds, then it got less again, but it was still there. And it kept floating, floating, and this took about two minutes after it arrived at the shed down there. It disappeared behind the shed so it was between the shed and the slope of Woodborough Hill. And it never appeared again."

"And the next day, was there any new formation in the vicinity?"

"No, but it was floating over a field of ripe oilseed rape nearly ready for harvest. It was very mature and ripe and about five feet high. It's very brittle and very thick and very painful when you walk through it. It's like little sticks. You can't walk through it because it's going to hurt you so much! The next day I went looking there to see how the situation was and there was no pathway, no way you could walk through the field. And I never saw a thing like that before and I never saw it again."

"And this was close to the East Field in 1996?"

"Yes, just on the opposite side of the road. When you drive from Alton Barnes toward West Stowell, you have the East Field on the left and Woodborough Hill to the right. And just passing Woodborough Hill and go a little bit further to the right, before you get to Tawsmead

Copse, there is a field. And that field was oilseed rape on the opposite side of the East Field."

"What formation was in the East Field then?"

"It was the DNA."

Or sines and cosines. That's what a mathematician suggested to me. Sines and cosines are depicted as a wave pattern that is used in trigonometry to determine unknown angles and distances on both flat surfaces and in three-dimensional space.

East Field formation in Alton Barnes, Wiltshire, discovered at 6:30 AM *on June 17, 1996. There were eighty-nine circles and the length was 648 feet. Photograph © 1996 by Steve Alexander.*

Back to 1999, a pictogram reminiscent of the early 1990s formations was found on June 23 stretched out 426 feet in a Stanton St. Bernard field not far from Alton Barnes. Investigators found many stretched and reoriented growth nodes and expulsion cavities associated with short, intense microwave exposure that Dr. Levengood has simulated in his laboratory. (See Chapter 8.)

Stanton St. Bernard pictogram in wheat, discovered June 23, 1999.
Aerial videograph © 1999 by Peter R. Sorensen.

Nearly two months after this long pictogram appeared, two more formations were found in the same field on August 14, 1999. A flat spiral emerged across the road from the East Field. Not far from the spiral was a second formation referred to as the "Celtic Knot." Peter Sorensen, I and others visited the field in the afternoon on August 15.

Spiral discovered in Stanton St. Bernard, Wiltshire, wheat near Celtic Knot on August 14, 1999, in same field as the earlier June 23 pictogram. Aerial videograph © 1999 by Peter R. Sorensen.

As we first approached the spiral, the downed wheat formed a narrow corridor about four feet wide that wound several hundred feet into a beautiful coiled center from which other wheat plants stood upright, almost like a crown, at the end of the flattened corridor. Along the edges of the spiral path, we found lengthened and reoriented growth nodes.

Then we moved into the tramline to walk about one hundred yards to a "Celtic Knot" made of wheat stems laid down in many layers, some crossing at ninety-degree angles, others lying 180 degrees from each other like hair parted down the middle. The complex pattern also contained lengthened and reoriented growth nodes and stems flattened to the ground without breaking or creasing. Peter Sorensen used his computer graphic tools to remove the tramlines in the Celtic Knot videograph so that its pattern could be seen more easily.

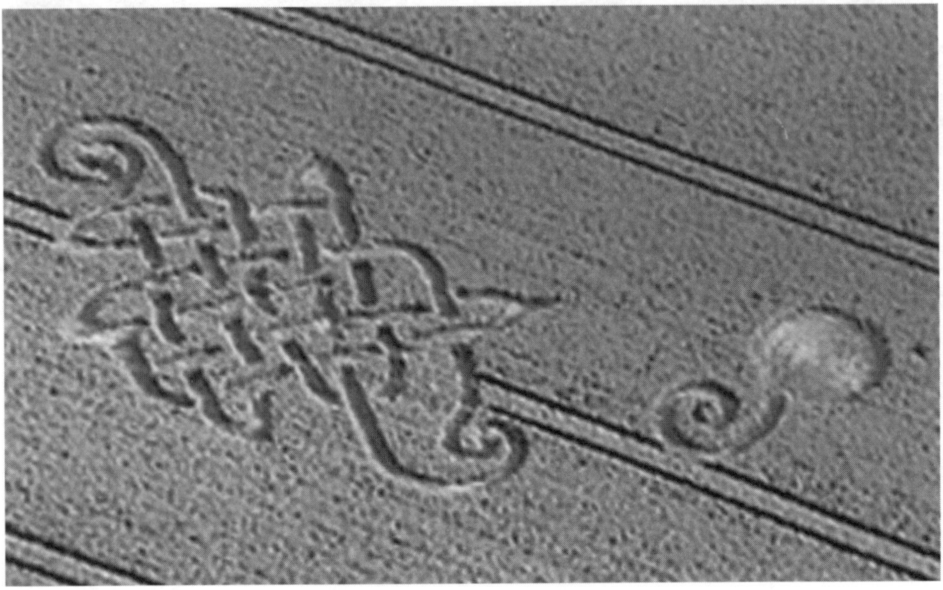

Celtic Knot aerial videograph retouched by photographer Peter Sorensen on computer to remove tramlines so pattern can be seen more easily. Aerial videobraph © 1999 by Peter R. Sorensen.

Lengthened and reoriented growth nodes along Spiral and Celtic Knot on August 15, 1999, in Stanton St. Bernard wheat field. Photograph © 1999 by author.

Wheat stems flat on the ground without cracks or creases in Celtic Knot on August 15, 1999, Stanton St. Bernard, Wiltshire. Photograph © 1999 by author.

Linda Moulton Howe on August 15, 1999, behind circle of standing wheat in Stanton St. Bernard spiral formation. Photograph © 1999 by author.

A few yards away was a large, flat circle with a curled tail and a small "grapeshot" circle outside. That circle and curled tail "signature" also had shown up before in a 1992 formation, Peter said.

Large wheat circle in curled tail "signature" next to "Celtic Knot" on August 15, 1999. in Stanton St. Bernard, Wiltshire. Photograph © 1999 by author.

As I moved down the tramline into the tail and approached the large circle, I began to feel physical pressure on the back of my neck and head. When I reached the circle and stepped inside, instantaneously I felt too light-headed to remain and stepped back out to restore my balance and try to better understand what was happening.

The pressure on the back of my neck and head in the big circle reminded me of what happened once when I was reporting about a Tesla Coil experiment in Denver, Colorado. When the scientist turned on the enormous electrostatic coil, I felt pushed back several feet by a wall of "jello" and nearly lost my balance. The scientist explained that electromagnetic and electrostatic fields affect some people more strongly than others.

Now I was in a crop formation in Stanton St. Bernard, Wiltshire, England, experiencing the same pressure. But what had created the field of energy out there? If a spinning plasma vortex with microwave and electrical energies had been there, as Dr. Levengood hypothesizes, why would there be a stronger residual effect in the signature circle? Is there something about different soils that might interact more strongly with magnetic fields, holding a residual frequency pattern of whatever energies had focused there for even a few seconds?

Close-up of large signature circle's perimeter reminiscent of a fluted piecrust, Stanton St. Bernard, Wiltshire, August 15, 1999. Photograph © by author.

The next morning, August 16, I left for London and Heathrow Airport to travel back to Philadelphia. Two weeks later I received an electronic message from Peter.

"I got an e-mail tip on September 1st about a new formation northwest of the Avebury stone circle and let Ulrich know so he could fly and videotape it. This is the latest formation I've ever known this late in the season. It isn't very special looking, except for a radial lay combined with a conventional clockwise lay in the big circle which makes it look like an eye — the pupil being the small radial lay in the center.

"This one had a 'signature' consisting of the Greek symbol, Pi. The allusion to circles is obvious.

"Aside from the radial lay I mentioned, there were very curious raised portions in the lay of several of the medium sized circles. These can be seen in the aerial shot as arcs and rings. The eightfold geometry is unusual, but not unheard of.

"A few days later, Francis and Charles Mallett told me that their older, 100% mechanical, single lens reflex camera refused to function in the formation. But the camera did work when they left. When they immediately returned to the formation, it failed again. I think they tried three times with the camera.

"More importantly, they reported that there was a thin, meandering line coming off the upper right part of the Pi symbol. The line was just two inches wide and the bends were just a few inches below the seed heads! (Similar to bent and darkened standing crop at the center of Cherhill 1999's signature.)

"I wanted to go back to video that, but I learned the farmer had cut the wheat that day. Unfortunately, no one I know took an aerial close-up of the Pi. I did take a pole shot of it, but I can't be sure if I can see the squiggly line."

This last English formation of 1999 was a circle with a double radial burst at the center with eight radiating arms, each consisting of two large circles. Each outer circle contained a small standing crescent and a very small grapeshot-sized circle outside the standing crescents. One hundred feet away was the Pi symbol, which is the 16th letter of the Greek alphabet and represents the ratio of the circumference to the diameter of a circle as 3.14159, a constant in a wide range of mathematical problems.

Andy Thomas reported in his *SC: The Bimonthly Journal of Crop Circles and Beyond*, "Each circle in the main design seemed to display a different example of astonishing swirl, most notably of the standing curled 'sausage' variety, first seen in one of the grapeshot at the 1997 Milk Hill fractal star. This closes a decade of wonders no one could have dreamt of fifteen years ago."

Avebury "Eightfold with Pi" formation in wheat discovered September 1, 1999, last major English formation of the year. Eight arms, seventeen circles and sixty-four little circles. Aerial videograph taken by Ulrich Kox and © 1999 by Ulrich Kox and Peter R. Sorensen.

Pole shot of Pi symbol. Photograph © 1999 by Peter R. Sorensen.

One of raised coils in the Eightfold Pi formation discovered September 1, 1999, in Avebury wheat field. Videograph © 1999 by Peter R. Sorensen.

As harvest closed the chapter on 1999 English crop formations, a new series began to appear in Canada. Paul Anderson, director of Circles Phenomenon Research Canada in Vancouver, reported twenty. The largest number, ten, in Saskatchewan were followed by four in Alberta, three in Ontario and one each in Quebec, Prince Edward Island and British Columbia. Some were associated with odd light activity over the fields as well.

By the first of July 2000, with so much of the summer yet to unfold, the phenomenon was more global than it had been in 1999. Thirty-six crop formations had been reported in England, eighteen in Germany, three in the Netherlands, two in Spain, two in Italy, two in the U.S., one in Russia, one in Canada and one in the Czech Republic.

The first German formation was discovered on May 1 in Burghasungen near Kassel. About twenty people — including Janet Ossebaard, Bert Janssen and Frank Laumen — were on a nightwatch near Zierenberg and Kassel, an area famous for crop circles, which have appeared there each year since 1998 on May 1.

I asked Michael Hesemann, publisher of *Magazine 2000* in Duesseldorf, Germany; author of *The Cosmic Connection*; and longtime student of ancient histories and religions, why Kassel would play a pivotal geographic role in Germany, as Avebury and Silbury Hill have in England. He responded in the following letter.

"In my opinion, it is no coincidence at all that Kassel became the German center for crop circle activities. It is indeed, next to Karlsruhe and Schleswig-Holstein (which are the other areas with frequent circles in our country), the land which comes closest to the sacred landscape design of the sanctuaries of Avebury and Stonehenge.

"Kassel is also the birthplace of Rosicrucianism in the 17th Century and the center of Masonic activities in the 18th Century. In 1614, the Kassel printer Wilhelm Wessel published two books which became the 'Bibles' of Rosicrucianism: the anonymous *General Reformation of the Whole World* and the famous *Fama Fraternitatis or the Discovery of the Commendable Order of the Rose Cross.* It was written by Johann Valentin Andreae, son of a Lutheran priest with the reputation of being an alchemist, born and living near Tuebingen in Wuerttemberg.

"One year later, Andreae published his second book *Confessio Fraternitatis or Confessuin of the Commendable Brotherhood of the Venerable Rose Cross, Addressed to the Intellectuals of Europe,* again in Kassel. The fact that Andreae spent all his life in Wuerttemberg, but published his books in Kassel/Hessen indicates that there was indeed a secret society already active at that time which coordinated the publication. With a very high probability, Andreae's friend, Rudolf August von Braunschweig-Wolfenbuettel, was another member of this Rosicrucian order. Its center of activities at that time was obviously already the city of Kassel.

"Mentioned for the first time in 913, Kassel obtained its town-rights by the end of the 12th Century when it became seat of a German chapter of the Knight Templars. Since 1277, it is the residence of the Counts of Hessen. At that time until 1848, Germany consisted of over a hundred independent states or counties ruled by Counts, Barons, Princes and Kings, united within the Holy Roman Empire of the German nation. In 1527, Hessen introduced the Lutheran Reformation as one of the first German states. The Counts of Hessen were proud of their French ancestry, sharing with the Kings of France a remote descent from Pippin, the Short, the father of Charlemagne.

"In 1717, the very year in which Freemasonry went public for the first time with the foundation of the Great Lodge of England in London, the Count of Hessen-Kassel, one of the first German Freemasons, completely changed the design of his residential city, Kassel. With the help of

Italian architects, he designed it as 'the city of man,' a model of the human energy body with its seven chakras. For this, he erected a network of fountains and monuments along a straight line, the cities' new axis, ending on top of a mountain. There, he erected an octagonal structure corresponding with the eight-petalled lotus of the Buddhist model of the Kundalini, crowned by a structure which is half obelisk and half pyramid, with a bronze statue of Hercules on top: the God-Man of the Greek/Roman mythology leaning on his club. Hercules stands for the perfect man. Seen from below, the whole structure resembles a Tibetan stupa. At his feet in the octagon, Count Hessen-Kassel placed the defeated Giant Encelados, symbol for the ungodly 'old man.'

"Interestingly, Carl of Hessen-Kassel, who ruled from 1767 on, became one of the closest friends and disciples of the mysterious alchemist, Comte du Saint Germain. Although Hessen preferred to reside in Gottorf near Schleswig, the capital of his land was Kassel. He hosted Saint Germain in the nearby castle of Louisenlund, where he allegedly died in 1784. In 1774, Carl of Hessen-Kassel was received as a Freemason in the lodge at Schleswig. In 1782, he organized the first national conference of Freemasons in Wilhelmsbad near Kassel.

"In conclusion, I can say that Kassel was Germany's most important center for Rosicrucianism with a ruling family deeply involved in alchemy, freemasonry and sacred landscape design. There is a direct connection between these ancient crafts and occult sciences and ancient Egypt as there is one between the stone temple of Avebury with its Silbury Hill pyramid and the land of the Pharaohs. Therefore, it does not seem to be a coincidence that right here in Kassel the circle makers found a second field of activity."

Michael Hesemann did not have a suggestion about why April 30 to May 1 should be the annual period the circle makers have chosen for their appearance near Kassel. But that repeating period is why the nightwatch group stayed out April 30, 2000, to watch the fields, hoping to see the creation of a formation. Janet Ossebaard told me by phone what happened.

"At 2 AM all of a sudden it became extremely dark. It was so weird! It was not like anything I've ever experienced before. We were sitting in the car on the front seat watching over this valley. There were no stars or moon because it was very cloudy. Dry, but cloudy. It was dark, but we could see the villages because you could see the orange streetlights and car lights. There was also a motorway running behind us that curved to our left, and we could see the cars. So there were constant lights.

"But suddenly at 2 AM, all those lights diminished. They became very dim gradually. It was not like they were switched off in electricity failure. Bit by bit, they became less visible over a matter of seconds. And Bert said, 'Hey, look at that!' And I said, 'Yeah, it's happening!' Then, it was completely pitch-black like I've never seen before. At the same time, there was no sound left. Normally there are many sounds at night. But there was no sound whatsoever! It was like some kind of dark lid came down over us."

"Did you get out of the car to see what it felt like outside?"

"No, we didn't. We had the windows open, so we could feel the atmosphere. There was no difference. We talked to each other. We were sort of whispering. It was so overwhelming and we had no idea what it was. The streetlights were gone. The villages were gone. The sound was gone. The cars on the motorway were gone. We were completely isolated from the rest of the world. And it felt very nice. It was a protected feeling.

"It didn't look like a mist, though, because at a certain moment there was a car driving right beneath where we were on a bit of a hill. And the car was pretty close. And we could see it very clearly and Bert said, "Hey, look at that! That is weird!" All of a sudden we see this car, but there is no fog. We could see the headlights clearly. Yet, you couldn't see anything else. As soon as the car was gone, everything was gone. It was like we were in a pitch-black universe, floating. Really weird!

"This happened from 2 AM to 2:15 AM. So, for about a quarter of an hour, it was pitch-black and then it lifted. It was like the blackness was pulled up from above because you could see the lights lowest in the valley became clear first. Then it came up and up until the highest lights on top of the hill we could see again. So it was like a big blanket came down from above, not like a mist. And then it was lifted back off again.

"Then it happened again at exactly 3 AM until 3:15 AM. But it was even more intense than the first time. Really intense. And at 3:15 AM it lifted again and I became very tired and had the feeling that it's happened. It's formed. It's there. There is nothing we can do. We didn't see it. There was no sound, no balls of light. But this is it. So, I can go to sleep now."

"And at first sunlight, did you wake up and look for a formation?"

"Yes, we all gathered, the Germans and us. Everybody talked about the strange blackness and everybody was convinced there was a formation. So we started driving around to look for it and we couldn't find anything. Eventually we decided to go home and went back to Holland, which is about six hours' drive. After we got there, Frank gave us a call. 'Hey, there's a formation. It was behind us!' I think the intelligence behind the crop formations plays with us, tests us, and I wonder what more will happen in 2000?"

Frank Laumen told me he could not find any evidence of human footsteps or broken plants. "The circles were perfect, and I don't know how that's possible for humans in oilseed rape that's like spiderwebs. For me, it's no question, this energy comes from above. Not from the ground. It comes from above."

First spring 2000 formation in flowering oilseed rape in Burghasungen near Kassell, Germany. Discovered after strange blackness dimmed lights twice as reported by a nightwatch group of about twenty people. Pole shot above and aerial below © 2000 by Frank Laumen.

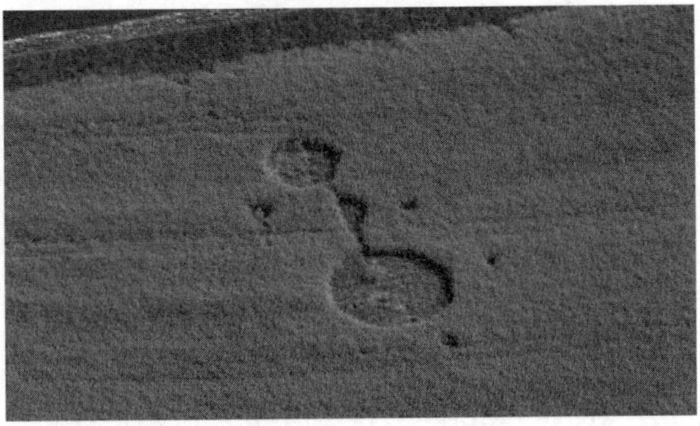

Back in Wiltshire on May 29, 2000, police received a call that a hole had opened up in Silbury Hill for the first time since 1776. That year the Duke of Northumberland hired Welsh miners to dig a seven foot square shaft a hundred feet down into the great, mysterious mound to look for buried treasure. He found nothing, the miners capped their dig back up and two hundred thirty years later in the rainy spring of 2000, the cap finally gave way. At least, that was the official explanation while archaeologists and engineers tried to figure out how to fill Silbury with hundreds of tons of new chalk. Helicopters were the easiest labor choice. Some local men went up the hill at night, got down inside the Silbury hole and said they found another eight foot square shaft angled off from the central hole. Whatever the origins of the shafts, such a difficult wound to repair after nearly six thousand years of Silbury's steady strength was an unnerving omen midway through the last year of the 20th century.

Another strange aspect of the story were eyewitnesses, including Frances Mallett, who saw "a ball of deep yellow light hanging in the sky approximately two hundred feet directly over Silbury Hill the night of June 1–2, 2000," reported by Charles and Frances Mallett at their C&F Phenomenon Research website at the cropcircleconnector.com. Then at 6:30 AM on June 2, 2000, the barley formation below was found at the West Kennett Long Barrow.

June 2, 2000, West Kennett Long Barrow, in barley, 150 feet in diameter. No one photographed an aerial before transformation into a second pattern twenty-four hours later. Here, Peter Sorensen used computer graphics software to reconstruct this first phase digitally. Digital image © 2000 by Peter R. Sorensen.

Peter, who saw the first formation from ground level, said the pattern consisted mainly of lines, which he tried to reconstruct on his computer in a before-and-after sequence. His aerial photograph below is the startling transformation found the next day on June 3.

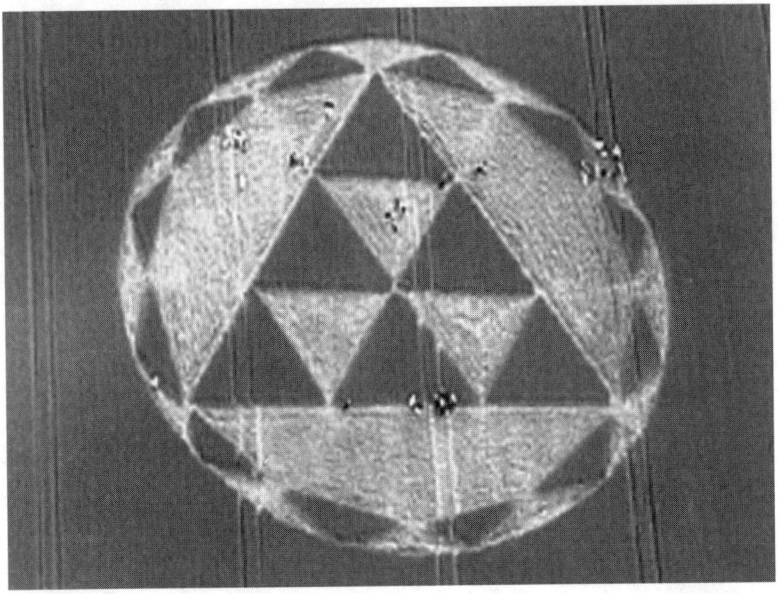

West Kennett Long Barrow formation on June 3, 2000, after transformation from the day before. Aerial videograph © 2000 by Peter R. Sorensen.

Tibetan Buddhist visual mantras come to mind, but nothing in *The Encyclopedia of Eastern Philosophy and Religion* exactly matches. Other people, concentrating on the central, lighter three triangles, recalled the old 1950s nuclear fallout shelter symbols that marked schools, hospitals and other areas in which the public could seek protection if atomic bombs were ever unleashed.

Peace, war, safety, danger, Horus and Seth.

"We are living in what the Greeks called the Kaitos *— the right time — for a 'metamorphosis of the Gods' ... This peculiarity of our time, which is certainly not of our conscious choosing, is the expression of the unconscious man within us who is changing."*

— Carl Jung, 1958

8

What Scientists Say

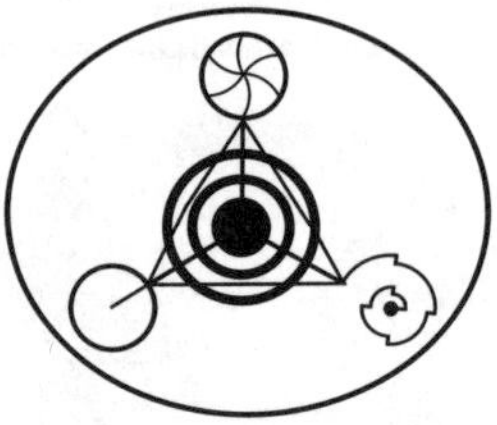

Biophysicist W. C. Levengood would agree with Frank Laumen that the energies that produce crop formations come from above. The formation that convinced him was Cherhill 1993, or at least the nearby field where Peter Sorensen dragged a magnet through a circle. The soil and wheat debris attracted to the magnet were covered with a thin coating of iron oxide. He already knew that microwave frequencies were involved because his laboratory tests with microwaves simulated the anomalies of enlarged cell wall pits, lengthened and reoriented growth nodes and expulsion cavities. All of these biophysical effects were caused by sudden and rapid heating of cytoplasmic constituents and water in the nodes, and with much more efficiency than soup is heated in a microwave oven.

He also suspected plasmas of positively charged ions and negatively charged electrons were the transport mechanisms for the microwaves and other complex electronic energies. Since the iron glaze was distributed like spray paint, over the soil and plants Dr. Levengood reasoned that the iron was in a liquid or semiliquid state rotating inside a spinning plasma before it was deposited by centrifugal force in most cases.

Many formations later, Dr. Levengood has demonstrated in his research reports that high concentrations of melted iron beads are often distributed in formation soils according to the laws of centrifugal force, and the empirical data have been shown to follow the the centrifugal force theory with correlation coefficients greater than 0.9.

Occasionally crop formations are not spiraled, but laid out radially, like spokes in a bicycle wheel. In such formations, unusual concentrations of iron particles have been found in the center where energies apparently focused straight downward. An example is the 1997 formation at Marion, New York, discussed in Chapter 3.

By the summer of 2000, Dr. Levengood had studied plants from at least three hundred formations in Israel, Australia, Canada, the United States, England and Holland. He found a repeating pattern of biochemical and biophysical changes in formation plants and magnetic anomalies in soils that lead him to conclude that what was interacting with crop formation plants and soil was "a spinning plasma vortex." But what was creating the vortices?

He has described his research in scholarly papers published by the international plant physiology journal *Physiologia Plantarum* and the American *Journal of Scientific Exploration.* So far, no one has refuted his hypothesis that spinning plasma vortices containing microwaves, magnetic fields and electrical fields produced by charge separation in plasma form the complex energy systems that create the crop formations. Dr. Levengood does not know the source of the vortices, but has hypothesized that the origination of the plasmas begins in the ionosphere or higher through complex interactions involving cosmic rays and streams of high-energy solar particles.

In a 1995 paper entitled "Semi-Molten Meteoric Iron Associated with a Crop Formation," published in the *Journal of Scientific Exploration* (Vol. 9, No. 2, pages 191–199), Dr. Levengood and his research colleague, John Burke, wrote: "Out of a four year investigation of plant tissues taken from over eighty crop formations, located within five countries, we report data from a single sample set which indicates a 'close encounter,' between the complex forces producing these flattened, circular type formations and meteoric material from the atmosphere.

"Both soil chunks and plant tissue taken from specific regions within a 1993 wheat crop formation in Cherhill, Wiltshire, England, exhibited a high degree of magnetic susceptibility which originated within adherent coatings consisting of the commingled iron oxides, hematite (Fe_2O_3) and magnetite (Fe_3O_4) fused into a heterogeneous mass, and having the properties of a 'magnetic glaze.' The makeup of this material appears to be of meteoric origin and apparently impacted the ground while in the semi-molten state. Presence of meteoric material adhering to both soil and plant tissues, casts considerable doubt on this being an artificially prepared or 'hoaxed' formation."

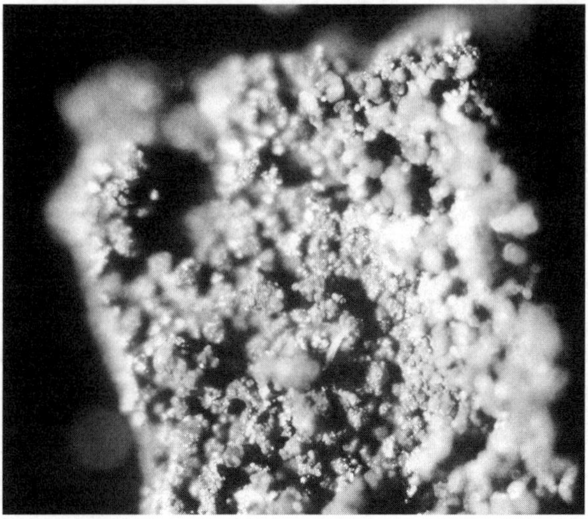

Heterogeneous mixture of the iron oxides hematite and magnetite on outer surface of wheat leaf found in Cherhill, Wiltshire, England, 1993. Magnification 40x. Photomicrograph © 1993 by W. C. Levengood.

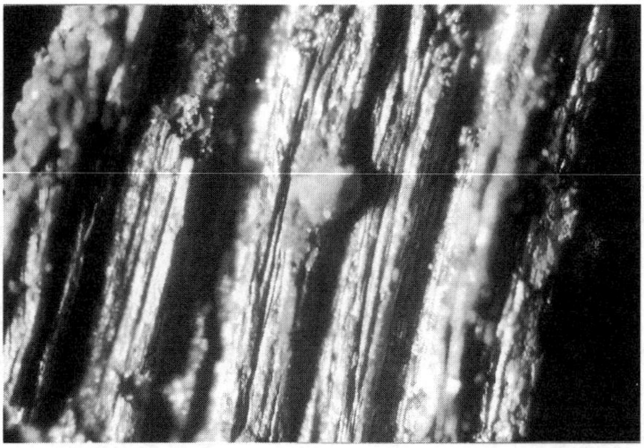

Flip side of wheat leaf: underside of same Cherhill wheat leaf that shows plant's epidermal cell structure imprinted in thin magnetic iron glaze. Magnification 40x. Photomicrograph © 1993 by W. C. Levengood.

The wheat leaf at Cherhill was not damaged by this molten iron deposit because of the Liedenfrost effect. Suddenly heated leaves can release steam through their stomata that creates a protective layer between the leaves and whatever warm substance touches them. In this case, it was the deposit of semimolten meteoric iron particles.

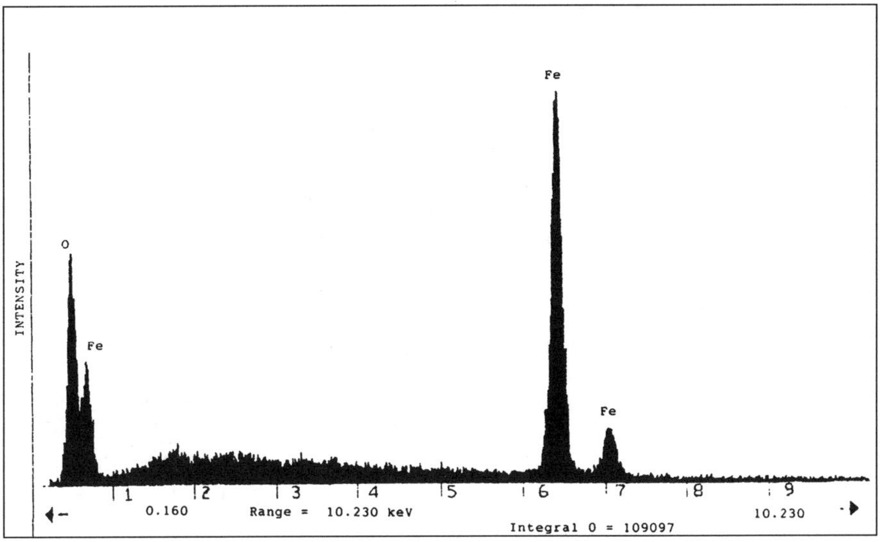

Energy dispersive spectroscopy (EDS) of soil in which magnetic glaze was found at Cherhill, Wiltshire, England, in 1993. Only peaks of iron and oxygen.

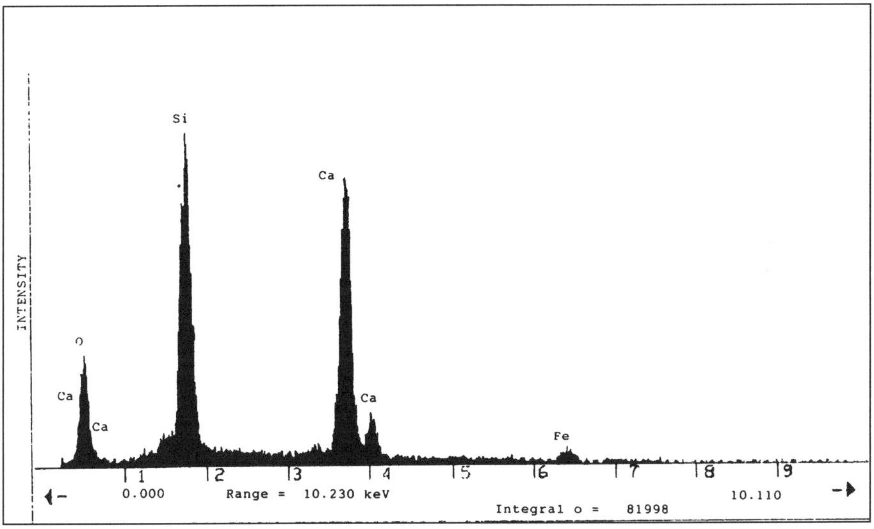

Energy dispersive spectroscopy (EDS) of normal control soil collected at Cherhill, Wiltshire, in 1993, showing calcium, silicon, oxygen and a small amount of iron.

Learning that the Perseid meteor shower in early August 1993 was more intense than usual, Dr. Levengood formulated a hypothesis about the origin of the iron magnetic glaze. He wrote, "In Cherhill, England, the meteoric dust was confined to separate smaller swirls within the larger swirls. Meteor trails are so highly ionized that they can be detected by radar (DeAngelis, 1988) and used to enhance radio communications. The structure of these ionized trails is still not well understood, but known to be turbulent (Gibbs, 1983). Plasma processes are commonly seen in the ionosphere within the aurora, many of which are spiral. In fact, the rays of the common aurora are now known to be tight vortices viewed from the side (Hallinan, 1981). While the full scale of an aurora is vastly larger than a 15 meter crop circle, the thickness is often little more than 100 meters.

"In plasmas, the magnetic pinch effect (of magnetic fields on plasmas) tends to concentrate the plasma into progressively tighter helical structures. As the overall width of the helix decreases, the local, induced magnetic field increases (Rose and Clark, 1960). Thus, if it were plasma vortices that impacted the English wheat field, the 0.5 meter sub-vortices would likely have had stronger magnetic fields than the 15 meter vortex of which it was a part. These smaller, high field strength regions could, by a process of plasma convection and chemical separation (Marklund, 1979), form discreet filaments containing the highly magnetic iron dust. Microwaves produced by the electrons rotating in their helical paths would heat any iron particles present and in air would also oxidize a portion to the hematite form. Internal heating may explain why the hematite-magnetite mixture was still semi-rigid on impact. Normally such droplets congeal in the air and reach the surface as independent spherules without plating the ground.

"Spiral auroras commonly contain arcs which have been seen to evolve into pairs of counterclockwise vortex sheets (arrays of vortices) that are never stable and never unwind. This compares well with the two, 15 meter counterclockwise spiraled areas in the wheat, with their attendant sub-vortices. Helical auroras originate in spiral electric fields of the magnetosphere and map down along geomagnetic field lines into the ionosphere (Hallinan, 1981)."

Dr. Levengood asked, "Is it possible that a pair of these downward directed, counterclockwise plasma vortices intersected and captured meteoric dust along the way, which in turn, was maintained or heated back to a semi-molten state by the microwaves of sub-vortices which carried it to the ground with its subsequent crop flattening energies? If so, this would indicate a heretofore *unknown phenomenon* of ionospheric plasma vortices descending to the Earth's surface. [Howe's emphasis.] Whatever the primary origin, the evidence clearly indicates that meteoric dust impacted the ground while in a semi-molten state and was confined to multiple, tight vortices within flattened wheat formations."

Since his Cherhill paper, Dr. Levengood has found high concentrations of magnetic iron in many other crop formations, occasionally more than concentrations 700 times higher than the normal 0.4 milligram of magnetite per gram of soil. Soil samples continue to confirm distribution by centrifugal spinning force.

Plasmas are distinctly different from solids, liquids and gases and are sometimes called "fourth states of matter." Well-known natural states of plasma on earth are lightning bolts, flames, auroras and gaseous neon lamps. Yet scientists have estimated that more than 99 percent of matter in the universe exists in the plasma state. Stars and our sun consist of plasma.

Plasmas are defined as a collection of electrically charged particles produced when atoms and molecules of matter are heated and release a soup of positively and negatively charged ions, electrons, neutral atoms and molecules, and photons. Physicists talk about high-energy and low-energy plasmas. High-energy plasmas are produced at high temperatures. Low-energy plasmas can form at room temperatures. Dr. Levengood says that low-energy plasmas occur when the electromagnetic charges become the primary determinant of molecular behavior. Low-energy plasmas, he thinks, are involved in crop circle formations.

The light energy of photons is continuously produced and absorbed in plasmas so that lightning, auroras, fire and lamps glow. Another phenomenon, known as "ball lightning," occurs as moving luminous spheres several inches in diameter near the ground during thunderstorms, which emit a hissing sound and distinct odor. However, ball lightning usually lasts only a few seconds and dies out suddenly, either silently or explosively. For more details and rare photographs of ball lightning, see "The Persistent Enigma of Ball Lightning" in this chapter.

Dr. Levengood thinks low-energy plasmas could explain the mysterious lights photographed in and around crop formations and sacred stone circles and dolmens. He thinks the reason could be electric and magnetic field anomalies at sacred sites that enable plasma creation.

This link was brought to my attention at Dr. Levengood's Michigan laboratory in May 2000. As he and I went through the images for this book, Dr. Levengood stopped at the Roundway photograph of Gordan Stewart sitting in the formation's center on August 1, 1999, with an unexpected arch of white light nearly obliterating his image. (See Chapter 4.) Dr.

Levengood showed me a very similar photograph taken by his research colleague, John Burke.

John told me what happened when he took the photograph. "In the fall of 1994, I took the enclosed photo of a dolmen rock chamber on Route 12 in Kent Cliffs, New York. It's important to stress these photos were by *flash*. I've gotten balls of light before at dolmen sites, but in flash photos *only*.

"The exciting part was that I had accidentally placed my blue backpack on the ground in front of the dolmen. Within an outside pocket of the pack was my electrostatic voltmeter and, as the photo shows, the mysterious arch of light rose from that spot. The meter is a standard scientific instrument which records electric charge in the air. This incident left the meter needle stuck on the maximum positive scale, something which never occurred before or since. It took six months for the accumulated electrostatic charge to bleed off and the meter became operational again."

For the reader's convenience, I have placed the Roundway photograph again below the Kent Cliffs photograph for comparison.

Flash photograph of rock dolmen mound in Kent Cliffs, New York. White knapsack on left in front of stones contained John Burke's electrostatic voltmeter that pegged and froze while he was across the road taking this photograph and the close-up that has the mysterious white arch. The knapsack is right below the emanation point of the arch. Photograph © 1994 by John Burke.

Flash photograph at Kent Cliffs, New York, rock dolmen mound in fall 1994 shows mysterious glowing arch anomaly with internal structure. Might be a low-energy plasma associated with intense electrostatic energy spot where researcher John Burke's electrostatic voltmeter needle froze on maximum positive scale. Photograph © 1994 by John Burke.

Flash photograph shows mysterious arch with internal structure that seems to spiral like a plasma. Not seen at the time. Gordan Stewart from Earls Barton, Northampton, England, is sitting at the center of the Roundway formation on Sunday, August 1, 1999. Photograph © 1999 by Bob Nicholas.

"This was not the only electromagnetic effect at this site. My battery-operated camera repeatedly failed to function there, although it would function properly when I drove ten minutes away to buy a new battery. I was impressed, but since I had seen such camera and electrostatic voltmeter failures at other ancient sites, I was not surprised. The reason I was studying these chambers, hundreds of which dot New England, was the fact that other scientists had repeatedly mapped magnetic anomalies at the spots chosen for some of these unexplained stone structures, and I had been able to confirm this pattern at other sites as well.

"American archaeologists do not know what to make of these structures, despite the fact that they are identical to hundreds of others in Western Europe and elsewhere and match the category professionally referred to as 'dolmen,' which Webster's Dictionary defines as 'any prehistoric megalithic structure consisting of two or more upright stones with a capstone, typically forming a chamber.' Radiocarbon dating in several New England locations has revealed construction dates anywhere from a few centuries ago back to at least 1,200 A.D. How could their builders have known that the earth's magnetic field was stronger or weaker in those sites without the benefit of scientific instruments to tell them so? As it turns out, more than one study shows that even modern people can frequently feel such differences under the right conditions. So whatever the purpose was for the Native American builders to construct mounds and dolmens, electromagnetic energy certainly played a central role."

I asked Dr. Levengood and John Burke why both photographs have a similar anomaly that was *not* visible at the time in either England or New York. They responded, "Low-energy plasmas are found throughout nature and by definition do not emit enough light on their own to glow sufficiently for the human eye to see. Though invisible, their existence can be confirmed with instruments, such as the electrostatic voltmeter which measured the site of the arch in the dolmen photograph at Kent Cliffs, New York.

"Another example of invisible plasma is the column of previously ionized air which a lightning bolt follows in a zigzag pattern. When the energy of a camera's flash is added to a weak, invisible plasma, that energy briefly moves the plasma's electrons in the outer shells of its molecules into a higher energy state. Those molecules almost instantly drop back down to their original energy state and emit added energy along the way in the form of light.

"A camera can record that light, but the speed at which it happens is usually, although not always, too fast for the human eye and brain to register. Thus, the photographer and people present see nothing, but the event is recorded on film or videotape.

"Lightning, for example, is a more powerful, longer-lived version of the same phenomenon. With lightning, instead of the flashbulb, it is the release of massive electrical charge through the initially invisible plasma column that produces the same rise and fall of energy levels. We see a lightning bolt because it is more powerful, longer in duration, and our eyes and brain can absorb the intense light.

"Recent work with lasers and high-intensity, noncoherent light sources has demonstrated that low-energy nonvisible plasmas can be readily stimulated to produce very intense pulses of photons which are very brief (in the nanosecond range) and peak in a different part of the spectrum (different color) than the stimulating energy. These studies clearly suggest that other light sources such as sunlight could produce high-intensity emission in organized low-energy plasmas. In fact, some well-organized plasmas are seen in photographs taken without a flash.

"With camera flash stimulation at other Native American sites, we have recorded many balls of plasma that did not show up on nonflash photographs taken at the same time. These were most common when the electrostatic charge in the air was high and measurable. In other cases, infrared photos without a flash have recorded balls of light which were not visible to the human eye at the time. We conclude that energy is present which registers as heat and electric charge, which can show up on flash photographs, but is not usually visible to the human eye.

"As with any natural phenomenon, these events are part of a spectrum. Unlike manufactured systems, nature usually works in a range of possibilities within parameters of natural law. At the outer edges of that spectrum are exceptions which occur less often. In one case, the light emitted after the plasma was stimulated by the camera's flash lasted long enough for the photographer to see. In more rare cases, the plasma appeared on photographs taken without a flash, but was not seen by the photographer. We hypothesize that this might be the result of extremely brief pulses of light emitted by the plasma itself — too briefly for the human eye and brain to register, but long enough for film and videotape to record."

In one of Dr. Levengood's English case files, he had photographs of stretched and reoriented growth nodes and expulsion cavities found in the wheat formation at Stanton St. Bernard on June 22, 1999. He stressed that "the energy causing node expansion appears to have microwave components whereas the energy inducing growth enhancement in the seeds has electrophoretic properties." Electrophoresis is the motion of charged particles, especially colloidal particles, through a relatively stationary liquid under the influence of an applied electrical field.

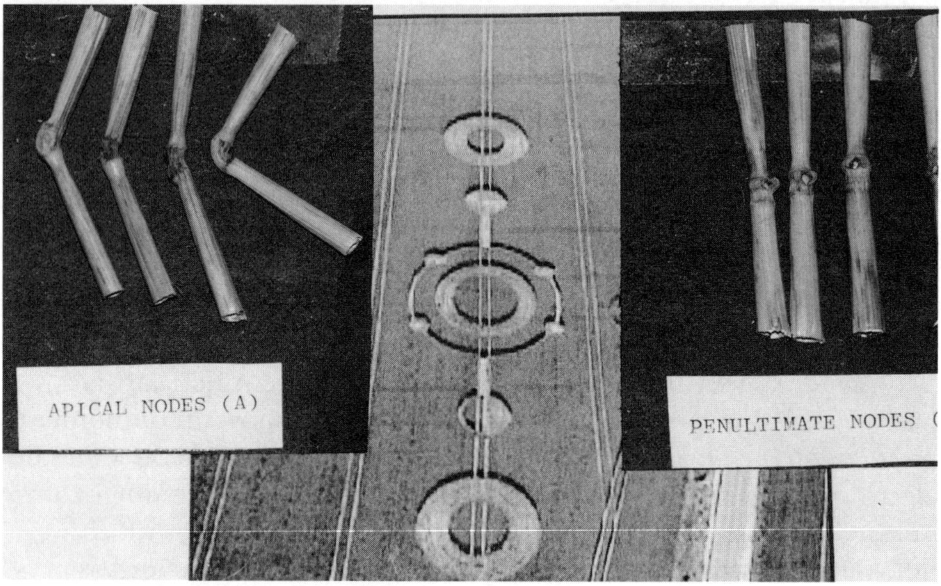

Stanton St. Bernard, Wiltshire, formation discovered June 22, 1999, in wheat field. Apical upper nodes were lengthened and reoriented. Penultimate nodes farther down the stems had expulsion cavities. Both effects have been simulated by Dr. Levengood by putting control plants in microwave ovens for a few seconds. Photograph © 1999 by W. C. Levengood.

"In fact," Dr. Levengood added, "in laboratory experimentation, we have successfully simulated the growth enhancement effects. John and I have found that when there is a drought, the plants grown from our experimental seeds don't wilt as much or as fast. The plants are resistant to drought. They can also last longer without light. The biochemistry of the plants has been changed to a much higher efficiency level."

I asked him, "What changed to make them such super plants?"

"The application of the plasma energy that we call 'ion-electron avalanches' stimulates the plants to produce a different biochemical system. We have found through chemical tests at outside laboratories that our plants have much higher levels of antioxidants. I have a theory about how these ion electrons form and interact with the plants."

"But what is it about the ion-electron avalanche from the plasma that would affect plant cell metabolism?"

"Let's take seeds, because we've got them worked out to a much finer degree. Seeds are living organisms. When the pulsing electric energies hit the seeds, or the plasma itself interacts with the plants, free radicals are produced that are destructive. But the plants try to counter the free radicals by producing more antioxidants. To produce antioxidants in reaction to the tremendous pulse of free radicals during the crop formation creation goes way back to my studies of ion transport in stems of plants and the effect of electric fields on them. Those are my redox tests of plant cell metabolism. What we've found is new stuff. You can't go to textbooks to find this.

"Another interesting discovery we've made is that if seeds that have been hit by free radicals are planted right away, they are practically dead. But if we wait several months, the plants end up more vigorous."

"Because the antioxidants have been at work?"

"I think it's in the genetic programming of plants. It takes time to turn on the antioxidants to neutralize the free radicals from those ion-electron pulses. But once antioxidant production is overstimulated, we see this terrific vigor in growth, health and ability to sustain drought, lack of light and other stresses."

In a recent grass formation discovered in Madera, California, on March 27, 2000, Dr. Levengood performed his redox tests of plant metabolism in grass from the center of a circle and in grass sampled from the formation's edge.

"The redox test came out about what I expected," Dr. Levengood said. "The controls had the normal level of respiration, and the other two from the formation were anomalous. After those tests, I put all the plants back in a light, tight box, completely dark, for a week. After eight days, I decided I didn't need them for any more work and opened the box to throw them out. I could not believe what I was looking at. The control plants were yellow and brown, dying, as you would expect. But the formation plants were still green!"

On right is grass from edge of March 27, 2000 Madera, California formation; on left is control grass sampled outside formation. Both grass clumps had gone without light or water for eight days. Formation plants were still green and vigorous compared to yellow, dying, normal control grass. Photograph © 2000 by W. C. Levengood.

Dr. Levengood said that in his redox tests, the edge plants EG2 should be the healthiest because their respiration was least affected by free radicals. Again, his theory is that the ion avalanches from a spinning plasma provokes more antioxidants in certain parts of formations, which in turn makes those seeds and plants more resistant to stress over time.

What might determine which plants will get the right balance of ion avalanche that provokes more antioxidant production? Contributing factors could include temperatures inside and outside the vortices, variations in the earth's magnetic field or inside plasmas, weather conditions, gas motions inside the vortex, and variations in electric fields produced by charge separation in the plasma vortices. Electric field variations, Dr. Levengood thinks, influence embryo development and seedling growth.

Dr. Levengood also showed me photos he had taken of deformed wheat stems found in the controversial July 29, 1999, Avebury formation. He concluded that whatever caused the unusual deformities had to have interacted with the crop *two or three weeks before* the formation was created when the stems were growing. Further, the deformities that Dr. Levengood calls "somatic developmental abnormalities" were also found in a Whitefish, Montana, wheat field on August 6, 1999.

The Whitefish, Montana, formation was nearly identical to a pattern from there the year before, in September 1998, that provoked the *Whitefish Pilot* to publish a photo on the front page of its September 3, 1998, issue.

95th Year / 36th Issue

This crop circle is located in a field between Whitefish and Kalispell.

Whitefish Pilot, *September 3, 1998, Whitefish, Montana.*

Whitefish, Montana, wheat formation discovered August 6, 1999.
Aerial photograph © 1999 by Melody Watts.

The same ring divided into four quadrants that projected outward with trident "forks" beyond the ring perimeter closely resembles a Native American symbol for the sun. The fields in which the 1998 and 1999 formations emerged were only seven miles apart.

If it had not been for the sharp eye and perseverance of Whitefish resident Melody Watts, perhaps no one would ever have known there was a repeat formation. After Melody saw it from the road while driving, she contacted local media and airports about photographing the pattern, but no one was interested. So she hired a pilot herself and took photographs. She also went into the formation and told me in a radio interview that the crop in the narrow ring was divided in the direction the plants were laid down. "Half of the ring was going clockwise, and the other half right next to it was going in the opposite direction. It was amazing." That meant there were two opposite flows of wheat in the same large ring.

Melody sampled the ring formation and gathered normal controls for study by Dr. Levengood. Mixed in with the wheat plants were "rogue oats" growing from the previous season.

"Some of those plants really had messed-up oat heads," Dr. Levengood said, comparing the oat photographs with the wheat in both the 1999 Whitefish and Avebury formations.

Dr. Levengood pointed out, "See, here in this Whitefish wheat plant, the first node is only a couple of centimeters below the head. I have one also like that from Avebury. Normally there are ten to fifteen centimeters between the head and the first node, or apical node.

"Have you ever seen any of these deformities before?"

"No, I was amazed. I've seen abnormalities in the seed heads, but not in the stems."

"What do you think happened to these plants?"

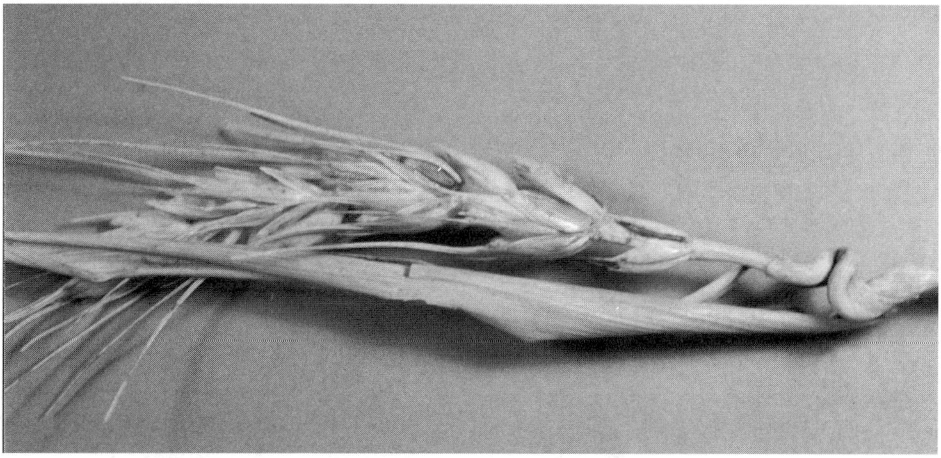

Wheat from Whitefish, Montana, formation found August 6, 1999. Somatic developmental abnormality in looping and twisted stem between first growth node and wheat head similar to formation in Avebury, Wiltshire, England. Photograph © 1999 by W. C. Levengood.

Wheat from Avebury, Wiltshire, found July 29, 1999. Somatic developmental abnormality in deformed looping and twisted stem between first growth node and wheat head similar to formation in Whitefish, Montana. Photograph © 1999 by W. C. Levengood.

"Sometimes when the energy blasts the plant, the size of the seed is reduced and the vigor is reduced. That means that what hit these plants probably occurred sometime after anthesis, which is after flowering. And so the energies injured the seeds and stem growth."

"What about the *Daily Mail*'s hoax claim?"

"I don't care whether the July 29 formation at Avebury was hoaxed or not. The fact is a spinning plasma energy hit that field probably two to three weeks before anthesis. That's the time when instead of the embryonic tissues being at a high metabolism, the tissue most active then was the stem tissue between the head and the first node. And the plasma energy just knocked the heck out of it and distorted it something awful."

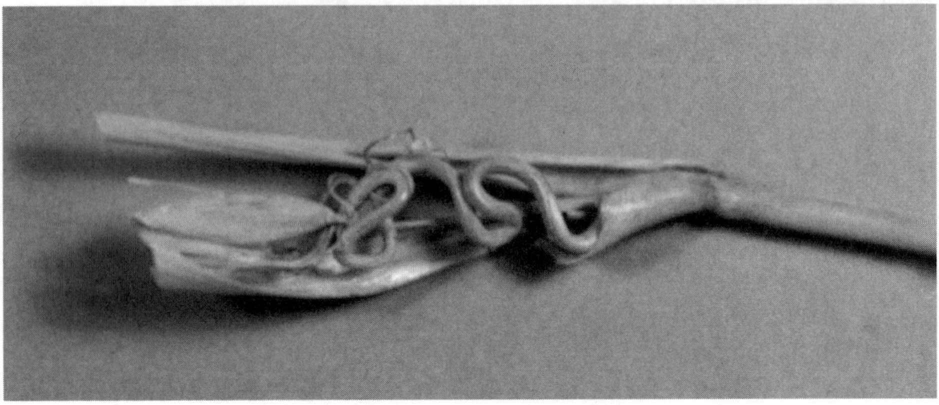

Whitefish, Montana, "rogue" oat plant from formation showing severe somatic developmental abnormality in looping and twisted stem between first growth node and seed head. Photograph © 1999 by W. C. Levengood.

"So the implication for these abnormalities to be in Avebury and Whitefish is that the plasma containing the microwave energies had to come down into the fields during the active growth of the stems?"

"Yes, it had to be then because when you germinate the seeds, they are normal."

"How long would that have been before the Avebury formation was discovered on July 29 and the Whitefish, Montana, formation was discovered on August 6?"

"I'm just guessing. I don't know if it would be three weeks, but at least two weeks before the formations were discovered. What is interesting is that the wheat heads developed normally, but the oat heads were highly deformed and they didn't have any seeds in them. But perhaps that's understandable because wheat and oats develop at different rates. The oats had deformity of the heads as well as the stems. In various parts of all living organisms, one part is growing faster one day and then another kind of tissue grows faster. And here we have two different species, oats and wheat, in two different locations. So the spiraling deformity occurred because the stem between the first growth node and the seed head was growing most rapidly when the plasma energies interacted with the field."

Some of these deformed plants also had expulsion cavities, which indicates there was a rapid heating of the node cell water and components that exploded outward.

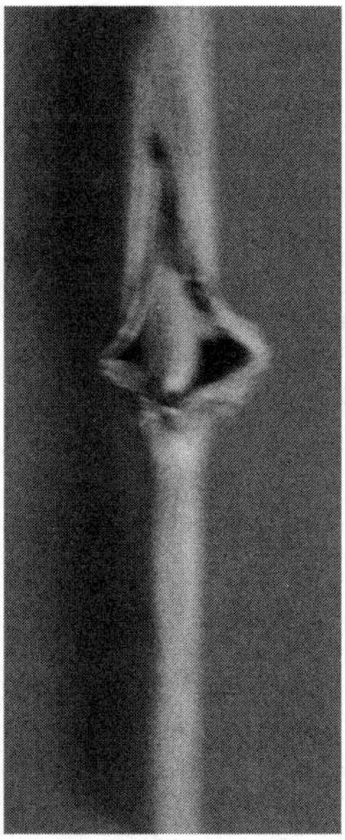

Expulsion cavities in stem deformities from Whitefish, Montana, formation discovered August 6, 1999. Photograph © 1999 by W. C. Levengood.

Summary of Biophysical and Biochemical Changes in Crop Formation Plants

a. *Lengthened and reoriented growth nodes.* These biophysical changes have also been simulated to a degree in normal cereal crops in laboratory exposure to various microwave frequencies for short durations of five to thirty seconds.

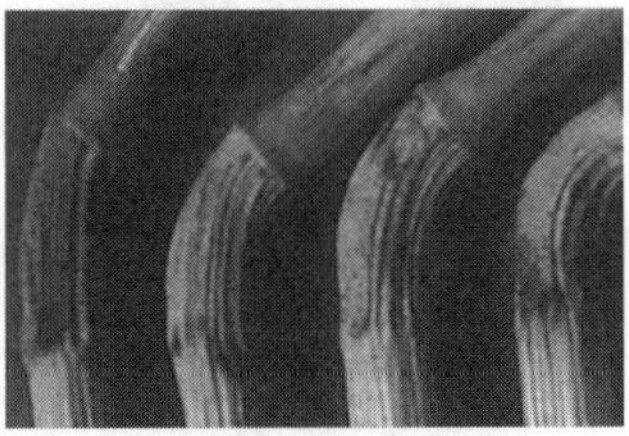

Blue Ball, Maryland, 1995, wheat.

b. *Expulsion cavities in growth nodes.* These form on the penultimate, or lower, nodes, which are much less elastic than the apical, or higher, nodes.

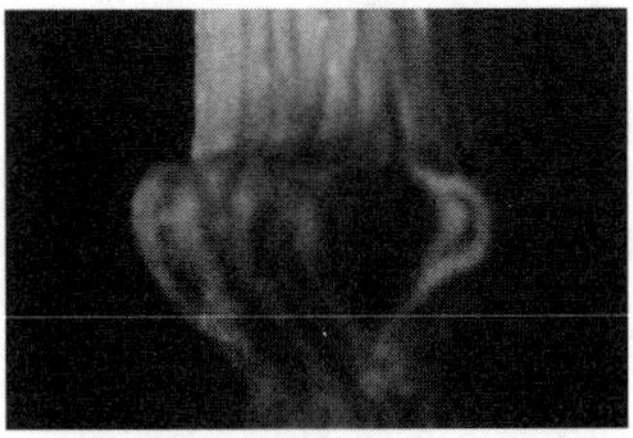

Logan, Utah, 1997, wheat.

c. *Deformed stems.* Rare "somatic developmental abnormalities," which Dr. Levengood links to microwave impact on growing plant cells.

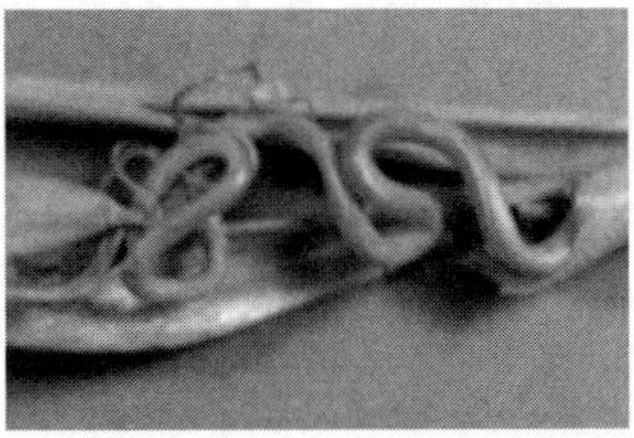

Whitefish, Montana, 1999, oats.

d. *Lack of seed development.* Formation seed heads on right compared to normal control heads on left. Likelihood that microwave frequencies altered developing seed embryos and stunted plant growth.

Alton Barnes, England, 1992, barley.

e. *Enhanced seed development and accelerated growth.* In some formations in which complex energies interact after seed development, plant growth is accelerated and next-generation seeds also show accelerated growth. Plants on left were grown from formation seeds. On right are normal control seeds for comparison.

Michigan Laboratory, 1992, corn. Accelerated growth on left.

f. *Enhanced vigor and resistance to light and water deprivation.* Dr. Levengood's hypothesis is that avalanches of ion-electrons from a spinning plasma vortex provoke plants in certain stages of growth to produce larger amounts of antioxidants than normal. Antioxidants would help plants resist deterioration when stressed by lack of water or light.

Madera, California, 2000, grass. Vigorous formation plants on right and normal controls on left after eight days without light and water.

g. *Reduction and Oxidation in Plant Cell Metabolism.* "Redox" measurements of mitochondria respiration in formation plants show stress under what Dr. Levengood hypothesizes is the impact of microwaves and ion-electron avalanches from the spinning plasma.

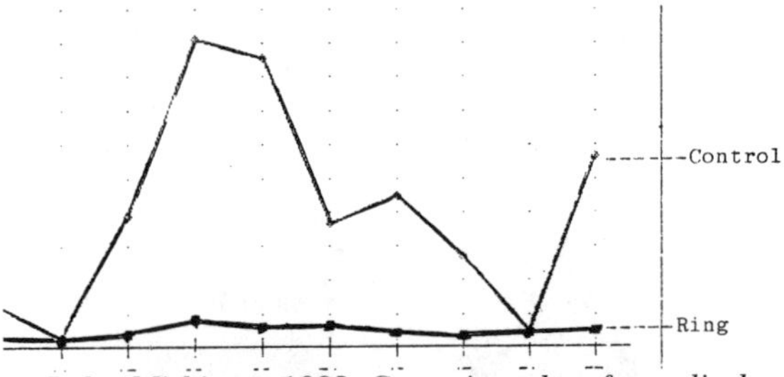

Grass Lake, Michigan, 1993. Grass ring where free radicals reduced.

h. *Compartmentalized Plasma Energies.* Dr. Levengood says, "One of the most ubiquitous characteristics of crop formations is the compartmentalization of energies into cells or quadrants acting totally independent of one another. The effects of the energies can change quite drastically within discreet, sharply defined regions inside a crop formation."

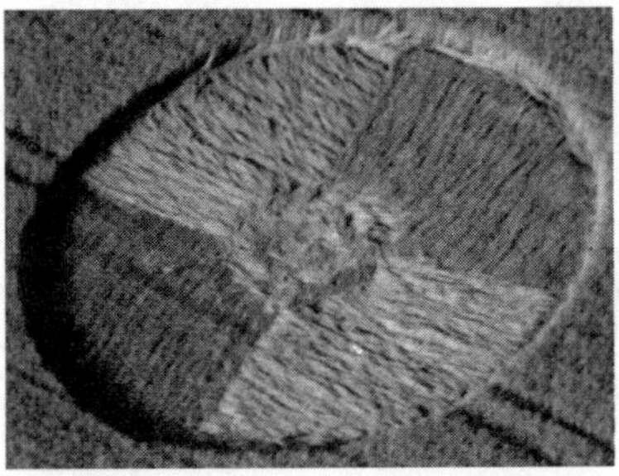

Wiltshire, England, 1989, wheat.

i. *Iron Magnetite Deposited on Plants.* Melted iron particles seem related to micrometeorite dust in the atmosphere, which can be drawn into spinning plasma vortices and deposited on plants.

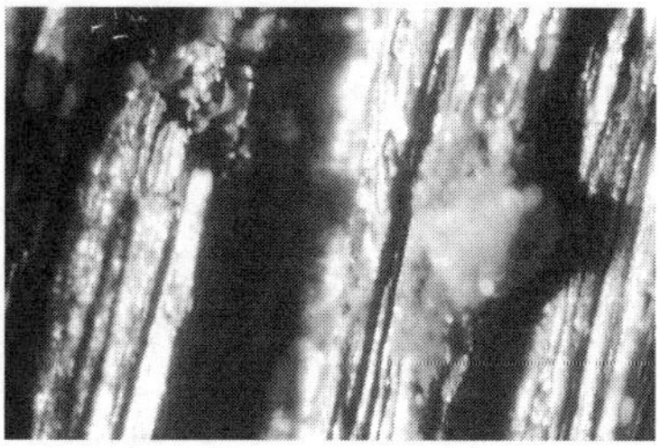

Cherhill, Wiltshire, England, 1993. Semimolten magnetic iron deposited on wheat leaf shows imprint of leaf's epidermal cell structure.

Summary of Physical Impacts on Crop Formation Soil

a. *Iron Magnetite Concentrations.* Significant increase (10 – 50 milligrams/gram soil) in amount of melted iron magnetite´particles compared to normal magnetite percentage in soils (0.4 milligram/gram soil).

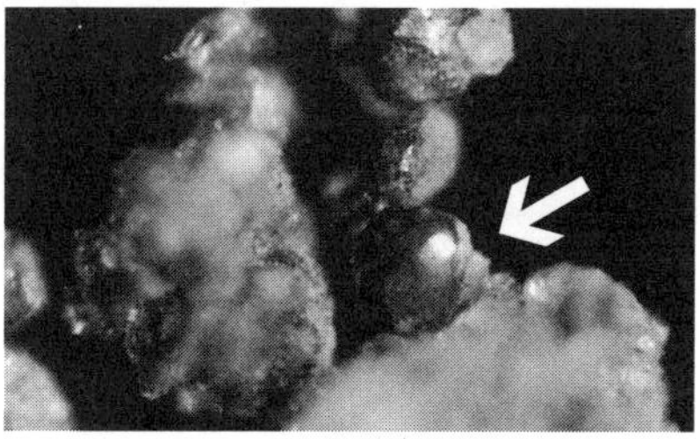

Cando, Saskatchewan, 1998. Melted bead of magnetite at center of soil particles in wheat formation.

b. *Glassy Substances.* Concentrations of micro-sized glassy spheres have been found in several formation centers.

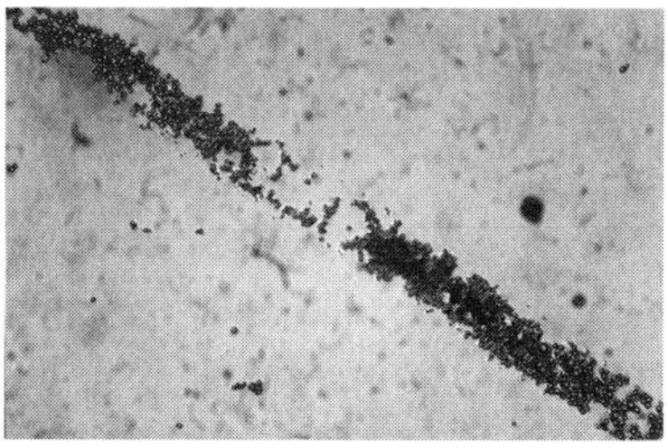

Logan, Utah, April 27, 1999, glassy microspheres.

c. *Unidentified Substances.* Unusual "pseudo crystals" have been found in hydrated soils from crop formations.

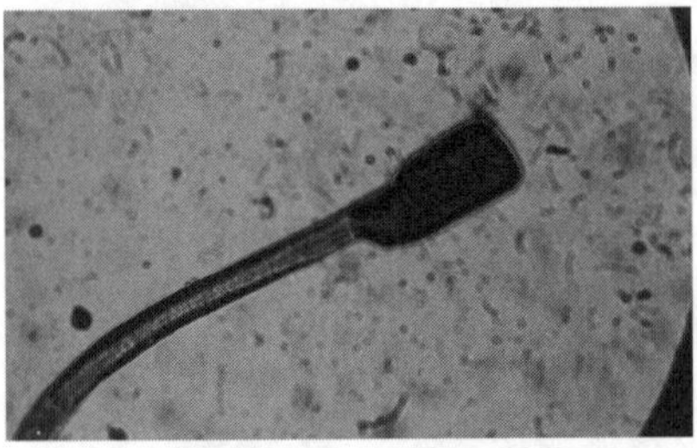

Oregon, 1999, unidentified "pseudo crystal."

d. *Centrifugal Force Distribution.* Iron particle distributions are often correlated with distance from the epicenter of a formation according to the physics of centrifugal force in a vortex.

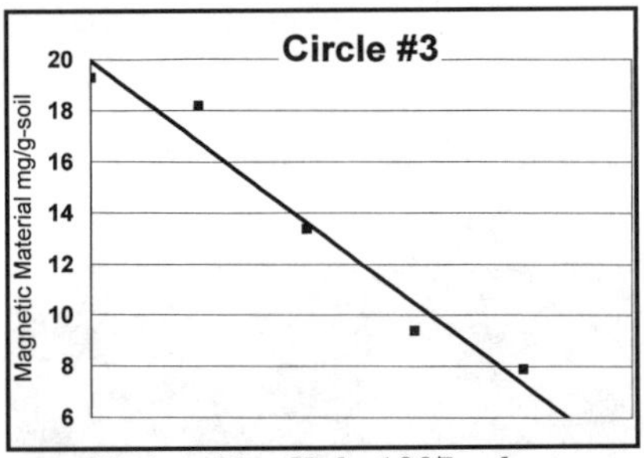

Marion, New York, 1997, wheat.

e. *Beer-Lambert Distribution.* Linear regression from crop formation data plotted according to the Beer-Lambert principle, relating the absorption of electromagnetic energy by matter to the distance from the energy source.

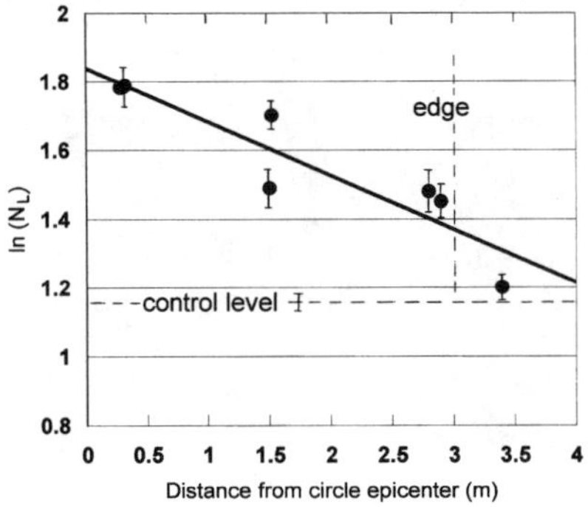

Devizes, Wiltshire, England, 1993, wheat.

Controlled Experiments in the Earth's Magnetosphere with Artificial Electron Beams

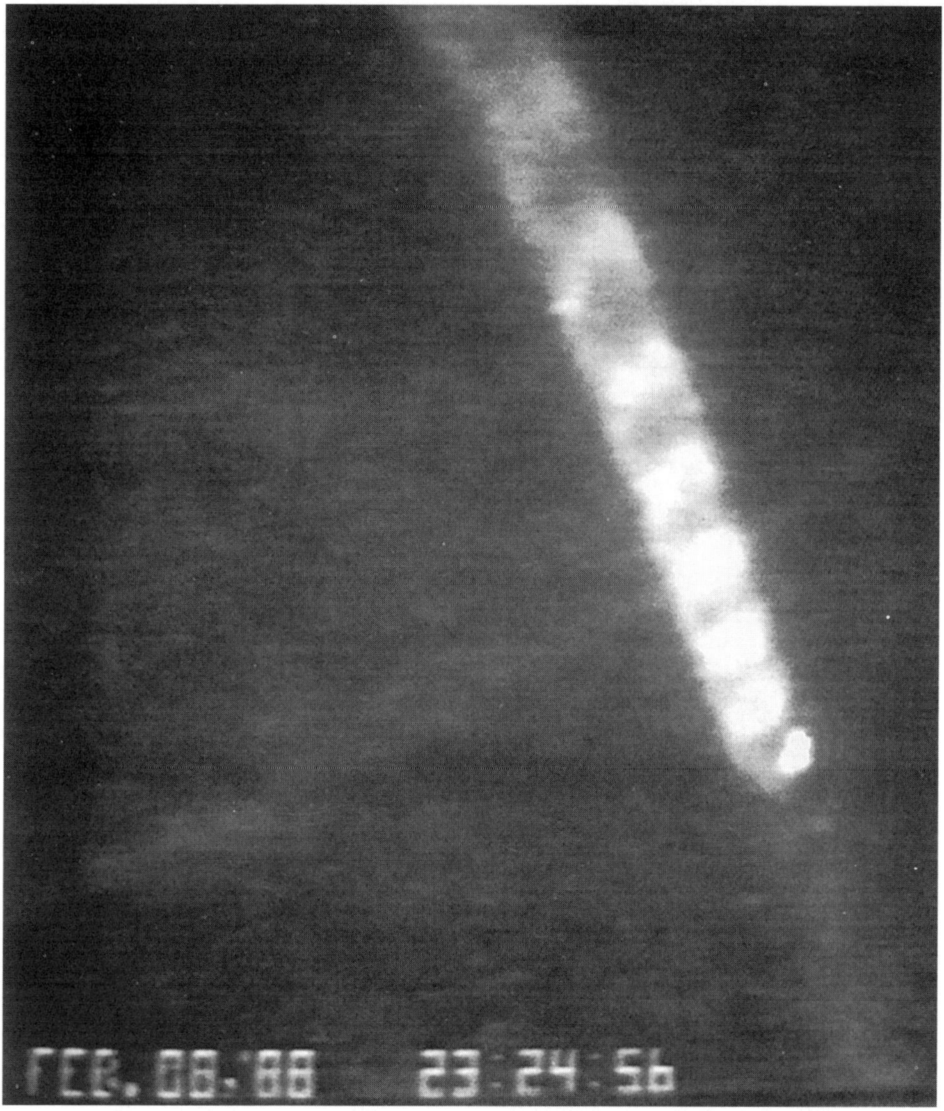

NASA ECHO 7 electron beam Larmor spiral revealed by the direct ionization of the dense neutral atmosphere near 100 kilometers. The image is that of a 36-keV (electron volts), 180-mA (milliamps) beam pulse of 150 ms (milliseconds) duration injected upward at a pitch angle of 130 degrees. Small bright glow is accelerator on the MAIN payload from which the beam originated.

For people who question the existence of spiral plasma vortices outside the confines of the plasma physics laboratory, it is valuable to consider the preceding NASA ECHO 7 photograph taken on February 9, 1988. The science team was led by J. R. Winckler, Ph.D., professor of physics and astronomy at the University of Minnesota, Minneapolis. The plasma research was funded by the Space Plasmas Division of NASA Headquarters, Washington, D.C., in their ECHO series of sounding rockets to research particle beams in space plasmas.

One of the purposes of ECHO 7 was to study electron beam interaction with the local auroral zone ionosphere at an altitude of about 100 kilometers. "The entire set of 93 downward injections from gun turn-on to reentry produced bright auroral streaks which were easily recognized and recorded by the TV camera," Dr. Winckler later wrote.

One of those NASA TV images shown on the previous page was featured on the cover of *EOS*, a journal of the *American Geophysical Union*, Vol. 70, No. 25, June 20, 1989. It clearly shows the classical motion of charged, glowing ions spiraling upward along one of the earth's magnetic field lines after the impact of the electron beam.

For comparison, here again is the New York dolmen arch that seems to have internal layers, perhaps spirals, as the Roundway photo shows.

Kent Cliffs, New York, rock dolmen, fall 1994. Glowing arch anomaly with internal structure. Might be a low-energy plasma charged by the camera's flash at the intense electrostatic energy spot that affected John Burke's electrostatic voltmeter. Photograph © 1994 by John Burke.

The Persistent Enigma of Ball Lightning

Often, as an investigative reporter out in the fields trying to understand and report about phenomena, I have been asked if mysterious lights associated with crop formations could be some form of ball lightning. For this book, I researched some scientific papers and journal articles to learn more about the illusive and controversial glowing balls. Some scientists argue there is no evidence that ball lightning exists. Yet other scientists, such as meteorologists and atmospheric physicists, have reported seeing such a glowing sphere.

Most ball lightning reports are associated directly with thunderstorms, a weather condition not present at the time of most mysterious light and sphere images in this book. However, I think it's valuable to compare images in previous chapters with two photographs of ball lightning I have been able to obtain that have been published in credible scientific literature.

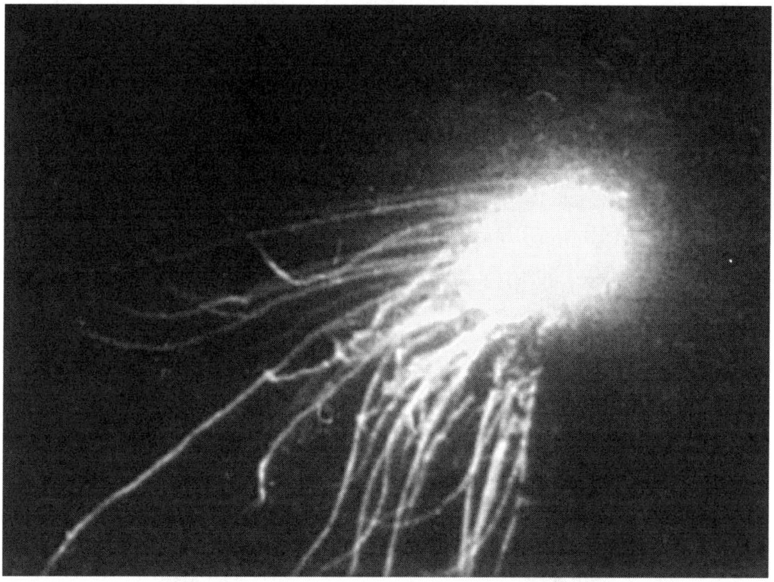

Photograph taken by eyewitness of round glowing head and trailing rays sometimes reported as characteristics of ball lightning. Originally published in 1951 in Naturwissenschaften, *Vol. 38, page 518.*

Another later 1980 paper in *Naturwissenschaften*, Vol. 67, pages 332–337 was written by physicist Stanley Singer, Ph.D. He said, "Ball lightning remains a mystery despite consideration extending over one and one-half centuries by a number of distinguished scientists. The fireball is observed in thunderstorms with intense electrical activity. The theoretical problem is that of accounting for a spherical structure which maintains its identity while moving freely in the air for some seconds and for a radiation process which continues over the same length of time."

Dr. Singer summarizes rare ball lightning as white, yellow, blue, green, red or orange in color with an average diameter of about twelve inches. It floats in the air for a few seconds and then disappears either silently or with a loud explosion. Some people have touched the glowing balls with opposite results — some say there was a surprising lack of warmth while others have said there was intense heat. "A crackling sound and odors such as those associated with an electrical discharge are noted."

Plasma spheres are on the short list of theories about what ball lightning might be. One scientist, Carlheim-Gyllenskoeld, in 1905 suggested that an ionized rotating current in the atmosphere might be responsible, reminiscent of Dr. Levengood's hypothesis that "a spinning plasma vortex of unknown origin containing microwave and ion-electronic pulses" is the system transferring energy to the crops, even if the source of the spinning plasma vortex is unknown. However, it is clear from this brief overview that ball lighting does not explain the long-lasting visible brightnesses nor the nonvisible spheres and lights associated with crop formation photographs and videotapes.

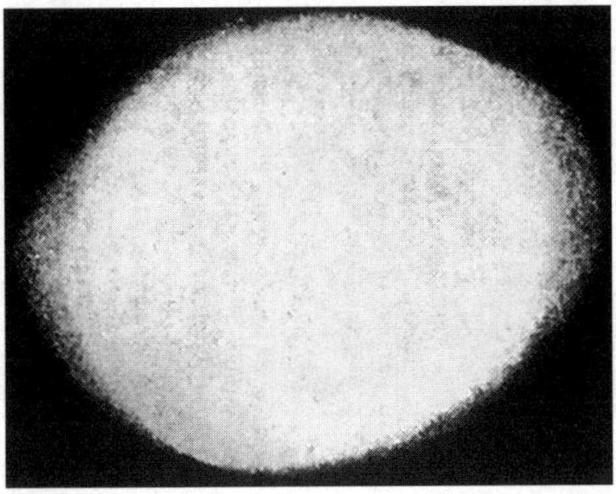

In 1934, an insurance appraiser was investigating lightning damage with several witnesses when he saw and photographed this glowing oval that lasted about ten seconds before disappearing. Photograph published by Feuerwehr-Verbands-Z, *Vol. 44, page 1, 1934.*

An Astronomer and Mathematician Looks at the Hoax Theory

Over the 1990s, mathematical analysis of the crop formation phenomenon has yielded insights about an intellectual profile involved in the creation of the crop patterns, which is either human-inspired or a completely independent intelligence whose intentions are not yet understood. Another scientist who took notice early on was Dr. Gerald Hawkins, astronomer and mathematician. Accounts of his work can be found under various headings, many published with peer review and listed in the bibliography. He began studying the geometries in 1990 after a career as chairman of the Astronomy Department at Boston University and research associate at the Harvard College Observatory. He co-authored with John B. White the 1965 book, *Stonehenge Decoded,* which hypothesized that Stonehenge marked sacred astronomical occasions such as solstices and equinoxes when the sun or moon was framed in the archways. Both Dr. Hawkins's and Dr. Levengood's early crop formation work were featured in chapter 1 of my 1994 book, *Glimpses of Other Realities, Volume I: Facts & Eyewitnesses.*

Dr. Hawkins also published in *Science News* and *Mathematics Teacher* his intriguing discoveries of new geometry theorems embedded in the famous 1991 Barbury Castle triangle, as well as repeating diatonic ratios in other geometric patterns such as the Litchfield 1995. Dr. Hawkins discovered five geometry theorems not found in Euclid or modern-day geometry texts. Theorems I through IV can be generated by Theorem V.

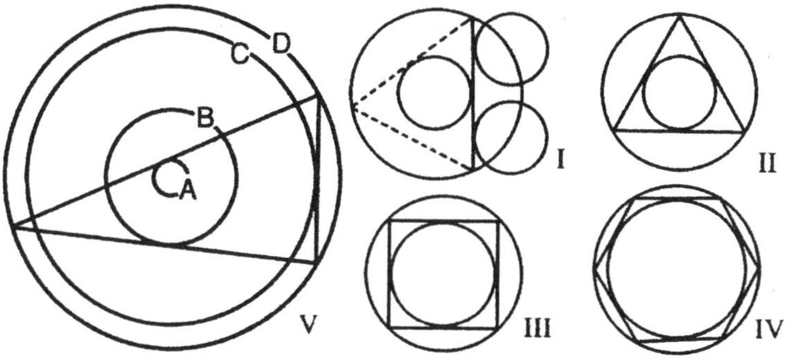

New geometry theorems I – V discovered by astronomer and mathematician Gerald Hawkins in the July 1988–1993 crop formations. Diagrams © 1995 by Gerald S. Hawkins.

I asked Dr. Hawkins if he would update his perspective on the crop formation mystery for this book. He sent me the following.

"Back in the early 1990s, I was in touch with the late Prof. Soli Zuckerman on circle topics, and friends and colleagues were asking me what was behind it all. They asked me because the crop circles were occurring near Stonehenge, the subject of my book *Stonehenge Decoded.* They wondered if there was some mysterious connection.

"Lord Zuckerman called them creations *of the dark,* a surprising claim from a former science adviser to the British government. There had been isolated reports in the distant past, but in sheer numbers and complexity, crop circles are a phenomenon of the end of the 20th Century. The media were all opting for hoaxers as the cause, so I decided to investigate the hoax theory.

"In dealing with crop circles, I have a fairly broad background, having earned degrees in astronomy, mathematics, physics and radio astronomy. Even more fortunately, I am my wife's harp tuner. As it turned out, it was harp tuning that was going to help me most! Julia has a folk harp with pure 8-note octaves like the white notes of the piano. She also plays a pedal harp where the black notes are added to make the 12-note octaves.

"My approach was to study the intellectual profile behind the patterns. The mechanics of how they are formed is a mystery, but the intellectual profile behind it all has turned out to be an even greater mystery. I was trying to prove the hoax theory, but I was not very successful in this. However, I made some important discoveries along the way. It has been a learning curve, particularly in the formative period, 1981–1993, in England.

"I found the circles contained musical ratios, new mathematics, and even a code which, by statistical analysis, has a high confidence level of intentionality.

"In the beginning, the circle sizes gave numbers which matched diatonic ratios. These are the step-ups in pitch of the white notes of the piano, or the lengths of the strings in a constant-tension folk harp. English church bells ring out with pure diatonic ratios and so does Big Ben with those famous London chimes.

"The diatonic ratios were established in 1981–1988 by the accurate field surveys of Colin Andrews and Pat Delgado. I found two simple rules in the crop formations. First, for satellite patterns and separated circles, you divided the large diameter by the smaller. Second, for concentric rings and shapes you divide the outer area by the inner. The confidence level in what I found was, by statistical analysis, better than 99%. So, the circle makers of that period knew about diatonic, or musical ratios!

"From 1988 onwards, we began to see geometries. I defined these as unembellished, rotationally symmetric diagrams based on four theorems. Circles were placed so that if a tangent were drawn between them, they made the theorem. But these creations were new mathematics — following on from where Euclid left off and not to be found in modern text books. Each one gave diatonic ratios corresponding to notes in the octave, C D E F G A B. The geometries are shown on page 275. For those who are visually-minded, the theorems are based on simple geometric shapes, with outer and inner concentric circles.

"For the mathematically-minded, the areas of the circles give whole number ratios. With regular polygons, only the triangle, square and hexagon give diatonic ratios and only these shapes were used by the circle makers in the formative period ending in 1993. In the triangle, the area of the outer circle divided by the inner is 4, and the annular ring gives 3. In the square, the ratio is 2 and the hexagon is 4/3.

"For the musically-minded, those numbers give C (3) in the third octave, G (2), C (2) and first octave F.

"Thus, three points on the circle makers' intellectual profile seemed to be visual, mathematical and musical. *Science News* editors were intrigued by the new math and ran an article on my findings in 1992. I had also found a fifth theorem not in Euclid or the text books, which was the starting point for the other four. I wondered if the unidentified hoaxers also knew this, so I kept it to myself. *Science News* readers wrote letters trying to guess it, but none succeeded. Then the fifty thousand mathematical readers of the *Mathematics Teacher* were thrown the challenge, but again none succeeded.

"Whether or not the circle designers read those magazines is a moot question because they did show knowledge of the fifth theorem in the so-called Celtic Torc at Litchfield in July 1995. After that, with the fifth theorem out in the open, *Science News* went ahead and published it. Then in 1999, the magazine migrated it for posterity onto their CD-ROM.

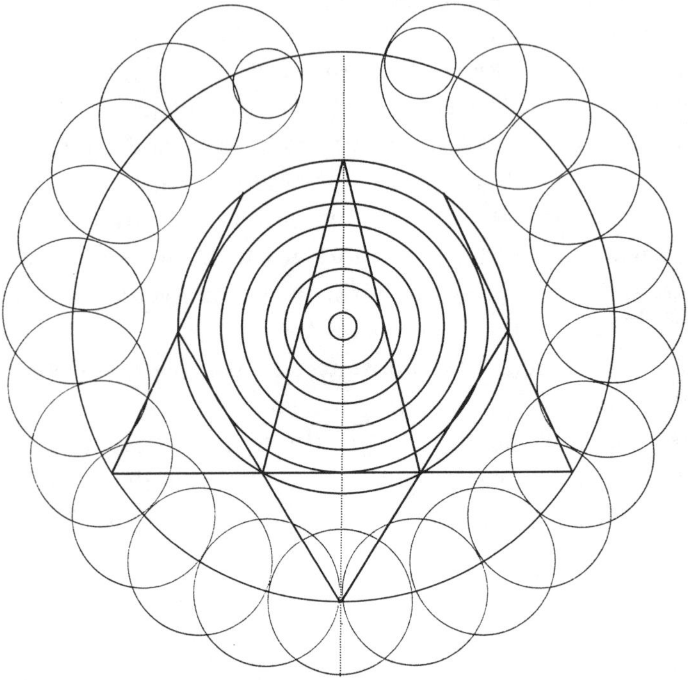

Theorem V geometry behind Litchfield, England wheat formation. Diagram © 1996 by Gerald S. Hawkins and Freddy Silva.

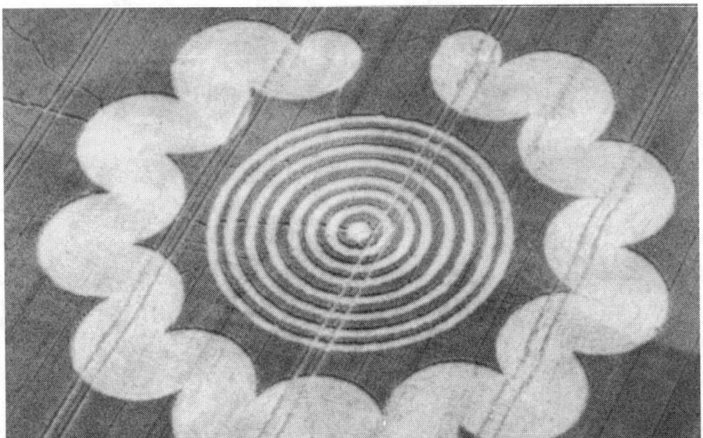

Aerial photograph of Litchfield, England wheat formation discovered July 6, 1995, about 200 feet in diameter, © 1995 by Lucy Pringle.

"I printed a summary in *Mathematics Teacher*, May 1998. The fifth theorem stated that when a triangle is drawn with the sides touching concentric circles, then as the triangle changes shape, the areas of the circles make the ratios of musical intervals. When the triangle is isosceles with two sides equal, it will generate the square and hexagon by rotation.

"The fifth theorem is unusual because the diagram moves, pulsing in and out, changing shape as on a computer screen. One could call this another point on the intellectual profile: 'Appreciation of mathematical niceties.'

"In the Litchfield formation, the designers used eight equally spaced rings. I proved that eight rings are needed before a diatonic triangle is possible and the base of this triangle touches ring seven and the sides ring two. It can hardly be a coincidence that tangents to ring seven intersect exactly on the outer ring, the Torc neckband. This 200-foot formation was accurately surveyed, and the geometry fit to within inches.

"I analyzed all geometries occurring in England from 1988 to 1993. All nine of them contained a pair of diatonic ratios. Linda Moulton Howe published some in *Glimpses of Other Realities, Volume I: Facts & Eyewitnesses.* Ray Cox gave an update in the *CCCS Circular* No. 35, 1999. Doug and Dave said they made circles 'for a laugh,' so we could not match the intellectual profile with those two hoaxers. Why were the unidentified circle makers giving musical notes? Was it a code?

"It was not unusual for musicians to encode messages in their compositions. Bach and Shostakovich did it using German keyboard notation, and Schumann went so far as to put the name of his paramour, Countess Abegg, into his very first Opus.

"The learning curve took us to English church organ keyboards and to a list. The first full octave in these church organs starts with note C, and the next keys have the letters DEFGAB. We can continue up the scales with the full alphabet, as shown in *Glimpses of Other Realities, Volume I: Facts & Eyewitnesses,* similar to the Boethean notation used around 500 A.D.

"Background information led me to see that the pairs of notes in the geometries spelled out the initials of seven persons on a short list: the first twenty-five presidents of the London Society for Psychical Research, the SPR. They were the great men of their day — William Crookes who pioneered the cathode ray tube by projecting a Maltese Cross; Charles Richet, Nobel laureate; Camille Flammarion, Legion of Honor astronomer; Oliver Lodge, inventor of radio (ultimately selling out to Marconi); John Strutt, also known as the physicist Lord

Rayleigh; Henry Sidgwick, philosopher; and Boyd Carpenter, the charismatic Bishop of the Edwardian era. Collectively they all searched for proof of life after death.

"These nine geometries gave the initials CR, BC, HS twice, JS, WC, CF, and OL twice. The corresponding geometries were: Corhampton '88; Winterbourne Stoke '89; Longwood Estate '90; Sompting '92; Fordham Place '90; Oliver's Castle '92; Etchilhampton '93; Uffington White Horse and Hogs Back Hill '93.

"What is the probability of this happening by chance? With 26 letters in the alphabet, there are 1/2 x 26 x 27 possible pairs of initials, AA, AB, AC and so on. So, the probability of any one random pair hitting any one of the 25 names on the list is 25/(1/2 x 26 x 27) = 0.071. There are nine geometries with nine hits. By Bernoulli's binomial theorem, the probability that this was intended is a very high 1.7 billion-to-one in favor.

"Take twenty-five names in sequence from any list, anywhere in the world, and the odds are it will not fit all those pairs of initials. I myself did not find the pairs and then search high and low for a fit. Nor was this like finding names by skipping through the letters of the Torah. The target list is very short. There is no room to wriggle.

"Scientists are persuaded by data that repeats. The code appeared in 1988–91 and was confirmed in the fields in 1992–1993. Ivars Peterson, author of the *Mathematical Tourist,* remarked to me how one could put aside arguments about statistics and look at the over-all credibility of the result. The list is not, say, a Little League team of South Succotash. It is a London list, set in the English countryside. English church bells are diatonic — no sharps or flats — and keyboard codes are part of the musical repertoire. Johann Bach memorialized his own name by putting B-A-C-H in the last fugue he wrote. For some reason, the circle makers were memorializing those men of letters.

"Even though my work is straight scientific research, it has been difficult to budge my colleagues. Perhaps they fear to tread into a subject beclouded with fringe speculations, or perhaps their background is not broad enough. Or perhaps, as one such colleague said to me, 'I don't understand it, and therefore, I'm not interested.'

"As a professional scientist, let it be known I am not invoking the spirits of those celebrated persons. No, I only point out that the names are being memorialized. In a way, what they stood for has been revisited.

"For me, there is a decided breakpoint after the 1993 crop circle season. Up until then, the patterns had been circles, rings, and rotational diagrams. All geometries fit the code. None were convincingly claimed to be the work of hoaxers. Nor did those few hoaxers show any interest in what I had found. After 1993, details of my work had been published and made available to the burgeoning groups. Simply to reproduce in wheat fields what had already been published would show us that these groups were up-to-date on reading the literature, but nothing more.

"Fortunately, no latter-day hoaxers could alter what took place in the formative period of 1981–1993. That data was safely sealed in a time capsule, beyond challenge. From my point of view, like the astronomy at Stonehenge which was set in stone, the mind behind the phenomenon was in the record, set in the wheat.

"Since 1993, there has been a tremendous amount of embellishment, but basic geometry is often a framework of the design. I have worked on a few formations that were brought to my attention as potentially significant. The formation at the West Kennett Long Barrow in August 1999 used the rule for satellite circles, though here it was satellites in the form of squares which repeated 'HS,' for Henry Sidgwick. The six yellow crescents at Barbury Castle in 1997 yielded, by Ptolemy's theorem of chords, the initials of Frederic Myers, an eighth president on the SPR short list, and pioneer in telepathy.

"At the end of a lecture I'm often asked: 'Well, Professor Hawkins, after ten years of research, what have you proved?'

"I say: 'Because there is an intellectual profile, I have proved crop circles can't be a natural phenomenon like whirlwinds, lightning or unguided plasmas. Furthermore, that profile is unique, touching the history of mathematics from Euclid to fractals and music from Pythagoras to English church bells.'

"OK, Dr. Hawkins, so what is your opinion?'

"My reply is open-minded: 'If all the crop patterns are made by hoaxers, then they should stop damaging our food supply. Clever math and musical codes should be published in the literature, not in the wheat fields.

"If the patterns are made by aliens arriving here in UFOs, then these extraterrestrials should stop giving us back our terrestrial information like the initials of famous but long-forgotten people of the Titanic era.

"If the phenomenon is transcendental, our culture is not currently prepared to face such a possibility; but if it is transcendental, then future society is in for a profound shock.'"

9

1999 Crop Formation Surveys by Andreas Mueller

There have been more than a thousand crop formations in southern England alone since 1989. No one person has surveyed them all, but several people have contributed hard work over the 1990s to surveys, diagrams and geometry analysis. Great credit should be given to Wolfgang Schindler, John Martineau, Michael Glickman, Paul Vigay, Peter Sorensen, Bert Janssen and Andreas Mueller. Their combined efforts would fill volumes and are an important archive of the complex range of designs and geometries.

In this book, however, my focus was largely 1999. The official total in England for the season was 159, according to Andreas Mueller and www.cropcircleconnector.com. Out of those, Andreas did several detailed diagrams of twenty-five major formations in Wiltshire, of which fifteen are reproduced in this chapter. He produced three pages for each to show the overall design, the complex directions in which the plants were laid down and the measurements of sections and overall dimensions. He also discovered complex lay patterns *beneath* the surface layer of plants such as a foundation in the five "pyramids" at the West Kennett Long Barrow and the fractal "ziggurat" at Silbury Hill.

Andreas has given me permission to share his 1999 surveys in this chapter and excerpts from an interview that I recorded with him at the Barge pub in August 1999.

Andreas Mueller was born in Saarbruecken, Germany, in 1976. He attended the Waldorf School, dedicated to anthroposophist Rudolph Steiner, from which he graduated, and in 1999 began his studies at the

School of Fine Arts in his hometown. He has examined more than two hundred crop formations since 1994, the year he began to focus on the mystery. His interest began with a school science project in which he could choose a subject to work on for a year. He chose crop circles, to study both their beauty and the provocative nature of their source. In 1994, Andreas founded I.C.C.A., The International Crop Circle Archive. I.C.C.A. has become one of the world's most complete and extensive databases on the subject of crop formations. It includes over three thousand different formation events that have been reported in at least forty-five-countries over the past five centuries.

One night at the Barge in Alton Barnes, Andreas talked to me about his perceptions of the crop mystery.

"When I first arrived here in Wiltshire, I realized that there are such beautiful patterns, and the whole phenomenon going on and most people don't realize it. In Germany, for example, the media doesn't cover the phenomenon.

"In June 1997 before I was going to England again, something strange happened late one evening when I sat on my balcony. It was about midnight. I glimpsed in the corner of my eye a light was moving very far away. And when I looked, I realized it was moving in my direction. I need to add that the house of my parents is in the countryside with meadows and horses. And so this little light came from a hill and rushed down through a valley where there are some wells and a creek.

"Then I looked at it hard with both eyes and it came zigzagging towards me very fast. It was not more than twenty meters away. It was in front of a very huge tree not far from our house, so I know how far away it was. And it was the size of a soccer ball. It was a sphere that resembled glass filled with illumination, like lime green light or green glowing smoke. It was very translucent. I watched it and followed it with my eyes. It was really there. I was stunned, amazed. But there was no interaction with me. It just continued its way through the valley very fast and disappeared.

"In 1994, which was my first year in England, we had a nightwatch on Knap Hill at the East Field. And at this time, there was an eye-shaped formation dubbed the 'Eye of Horus.' There were a bunch of people at Knap Hill and some had a very strong, professional searchlight. Around 11:30 PM, we saw inside the formation, probably in the center, a light started blinking, flashing, a white light, very bright, but very small. I would compare it to the thing Steve Alexander shot on the hill in 1991. And it was blinking a few times.

"Then immediately, people with the searchlight switched it on and this beam of light went over the field very fast from the searchlight. And we looked without binoculars, and in that second when the beam of light shone on the whole area, this blinking light starts moving fast over the whole field in a very high speed. Then it stopped at the edge of the crop and it was blinking three or four times again and then goes off.

"What we did then was scan the field with the light and binoculars and there was nobody in the field, nobody in the formation. After about five minutes or so, we switched off the light. Then, another light, or the same one, started again blinking three or four times, but on the opposite side of the field. After the three or four times blinking, it went off and we never saw it again.

"The only sense I have at this time is that there is a real phenomenon out there which is really beyond hoaxes and beyond little green men. And it's important to stay open because if you have a conclusion or some theory, then people create their own walls. Sometimes formations won't fit into a particular interpretation and people then say those formations are hoaxed. That's what researchers should never come to because that puts your idea or theory on top of the phenomenon itself. You end up judging the phenomenon by your personal view.

"Nature is creating similar patterns in snowflakes or in honeycomb hexagons, and there are so many geometric wonders you can find in nature. But that doesn't mean the crop circles are a natural phenomenon. I think each of us comes to the same question: What is behind nature? And then if you come to this point, you realize there is something behind nature also which seems to be intelligent.

"And what I basically feel myself about the crop circles, I don't know how they do it, but those shapes touch your soul and they change people's lives. We have people coming each year and they change their business. Before they were very in business for the money. And now they change it to what they want to do. It doesn't have to be anything with the crop circles. But just the way they want to have their life.

"I have seen over two hundred formations in the last two years. There are patterns that leave the usual geometry behind. For example, the Beltaine Wheel [ancient Celtic May Day celebration] which arrived May 4, 1998, at the West Kennett Long Barrow in fragile oilseed rape. It was a wheel made of thirty-three 'flames.' This shape you can reconstruct on paper by using a compass. But to get the final result as the formation was in the field, you have to rub out the lines which you have drawn. You are not able to do this in the field because when you flatten the part, you can never lift it back up as normal. Especially in oilseed rape. So this was very special to me. I work with geometries, and if you're used to geometries and you see these things happen before you on the paper, you realize this thing was not made in the usual way.

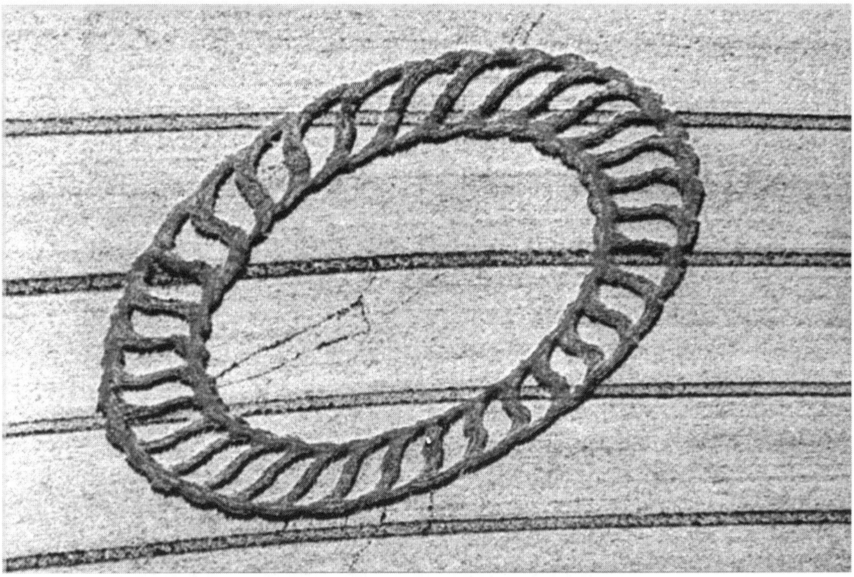

Thirty-three "flames" in a wheel called "Beltaine," discovered at the West Kennett Long Barrow on May 4, 1998, in canola (oilseed rape), 220 feet in diameter. Cameras and audiotape recorders failed inside it.
Photograph © 1998 by Lucy Pringle.

"Others did touch me on a different level. For example, the Triple Julia Set at Windmill Hill in 1996 where each of the 194 crop circles was made in a different way. Or when I know that a formation appeared in a very short time period, but I see this massive size that would be an overwhelming physical task. For example, the East Field formation in 1998 or the one in Roundway in 1999. They cover nearly six thousand square meters. This is a task!

"If I compare some events — like the dead flies at Cherhill in 1998 or the flattened porcupines [See *Glimpses of Other Realities, Vol. I*] in two Canadian circles in 1989 and 1992 — you have to accept that there were live creatures killed by the energy force that created the formation. But mostly, I feel it's nothing that is going to hurt live beings with intention."

Flies found dead with their tongues stuck to the wheat in the 1998 Cherhill Down formation. Other deaths in formations have included porcupines in Canada. Photograph © 1998 by Janet Ossebaard.

One of twisted 1998 Cherhill formation centers where flies were found dead and stuck to wheat heads. Photograph © 1998 by Janet Ossebaard.

"Andreas, if the crop formations are being done by a highly advanced intelligence, whether it's the Creator of the universe or an extraterrestrial or an angel or another dimensional or somebody from another universe or a time traveler — what could possibly provoke the energy, time and focus to make all of these formations around the world over at least the past ten earth-years?"

"One of the many ideas I could think of is that the formations really do something in your mind, even when you don't know it. That probably is the reason for the phenomenon — to switch a button in people's minds. And if a special number of people sees the formations and gets their own buttons switched, then the phenomenon or the source behind it will have enough people involved with this switched button. And then it just stops. And we don't know what the explanation for the crop circles is, but it happened to us without us knowing about it. This is a possibility. It would be very strange and very unsatisfactory. But it is one of the possibilities."

The following formations were all surveyed and hand-drawn by Andreas Mueller and excerpted in chronological order from his *Crop Circle Formations Diagram Trilogies 1999,* published by I.C.C.A., The International Crop Circle Archive, © 2000 by Andreas Mueller.

Honeystreet, Wiltshire, England

Discovered June 16, 1999, in wheat.

Diameter: 64 meters (210 feet)

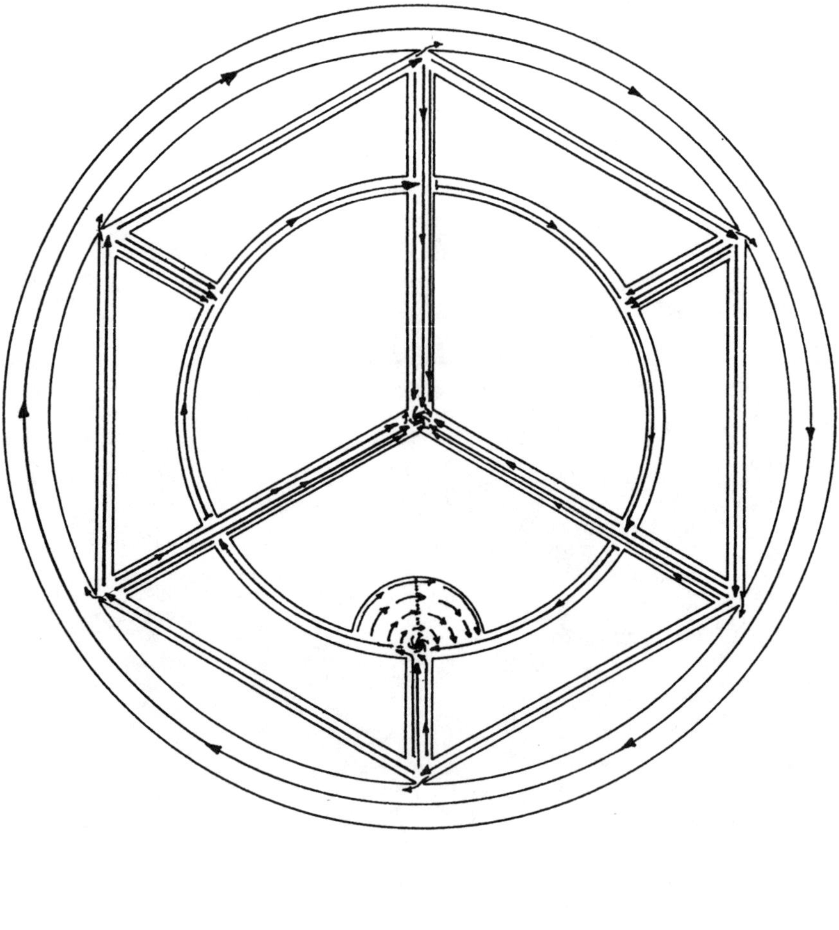

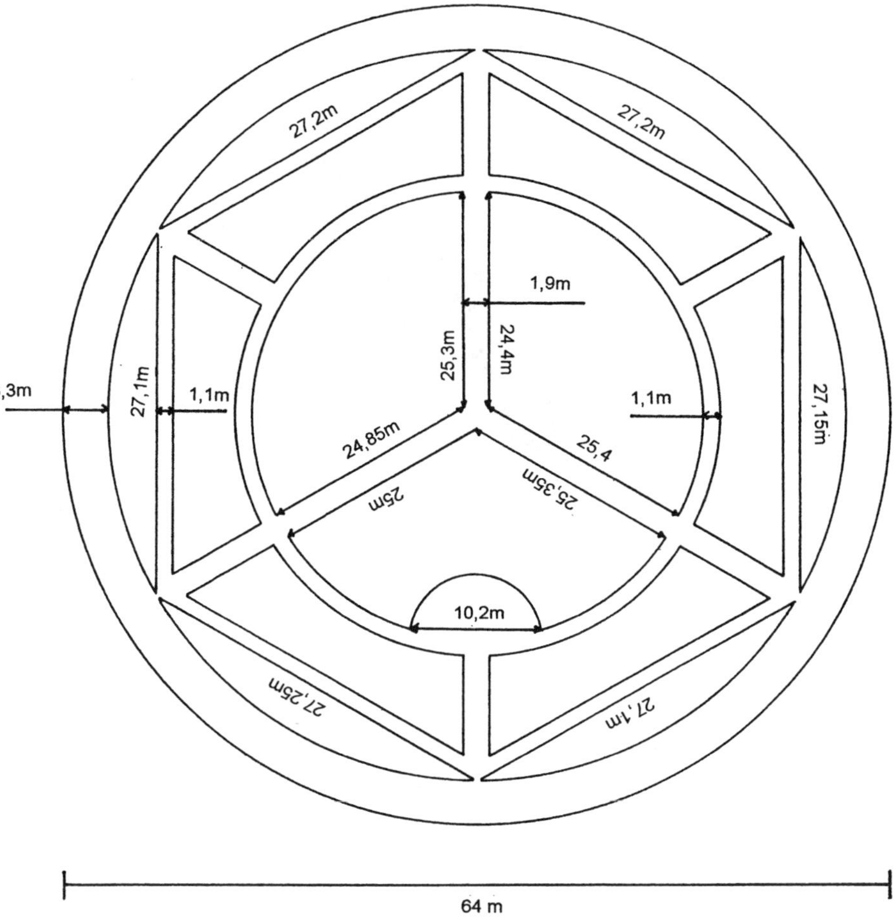
27,2m
27,2m
1,9m
25,3m
24,4m
3,3m
27,1m
1,1m
1,1m
27,15m
24,85m
25,4
25m
25,35m
10,2m
27,25m
27,1m
64 m

Stanton St. Bernard, Wiltshire, England

Discovered June 23, 1999, in wheat.

Length: 129.85 meters (426 feet)

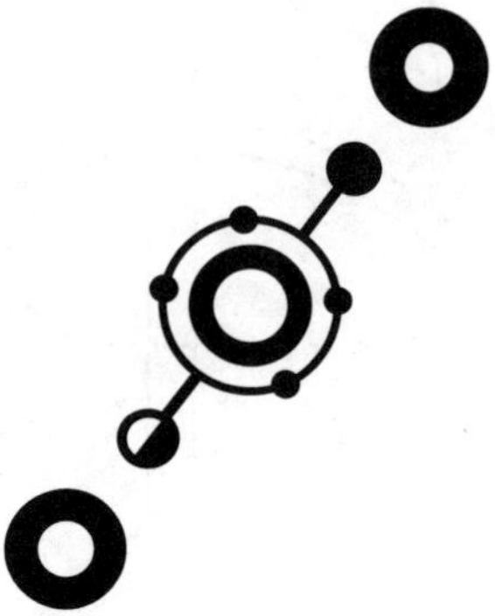

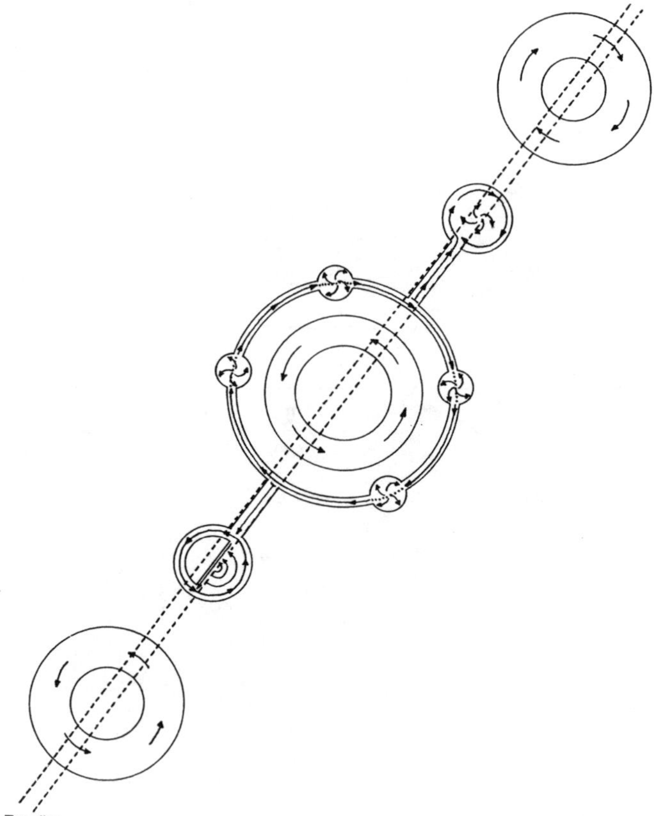

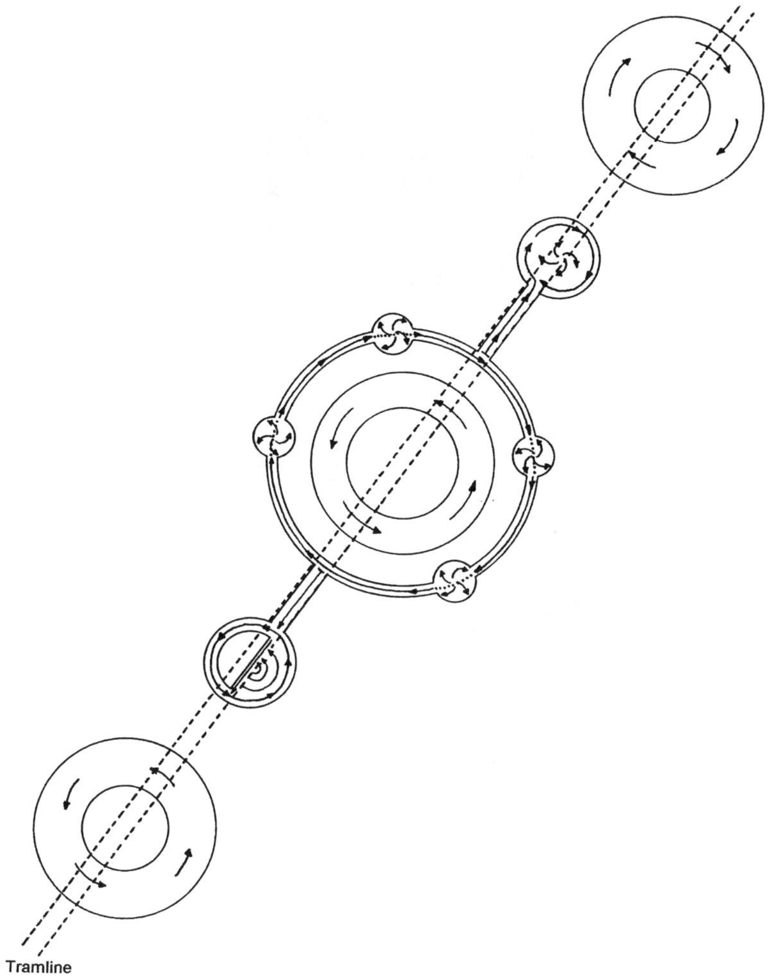
Tramline

Allington Down, Wiltshire, England

Discovered June 24, 1999, in wheat.

Diameter: 60.5 meters (199 feet)

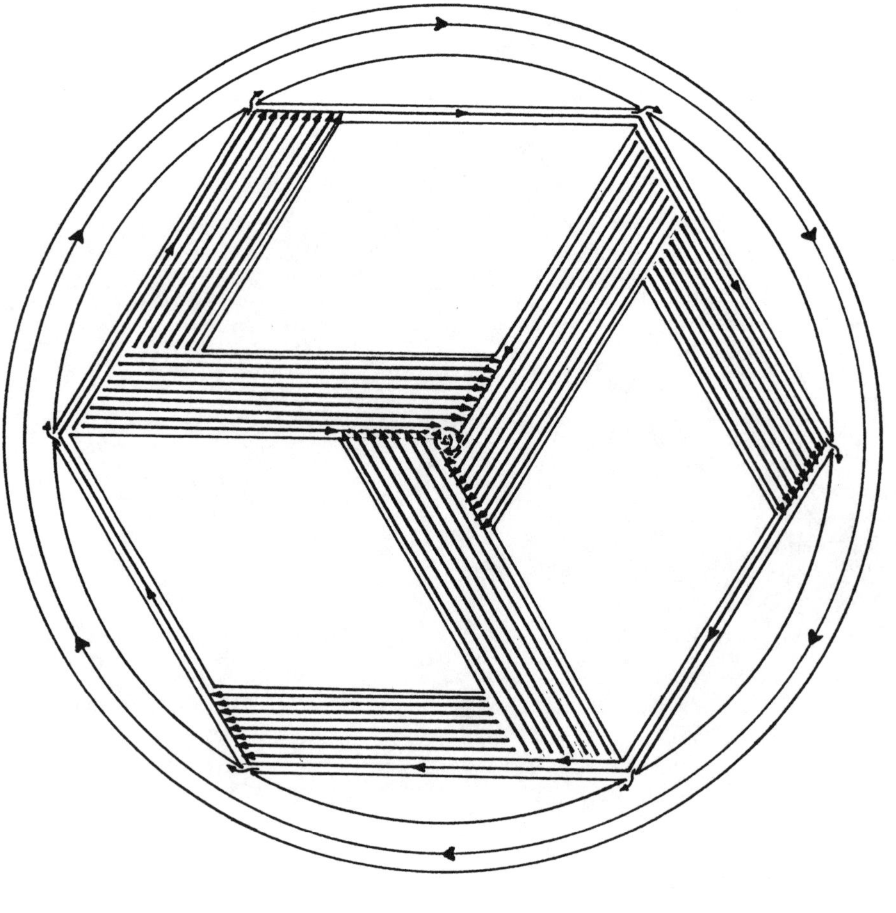

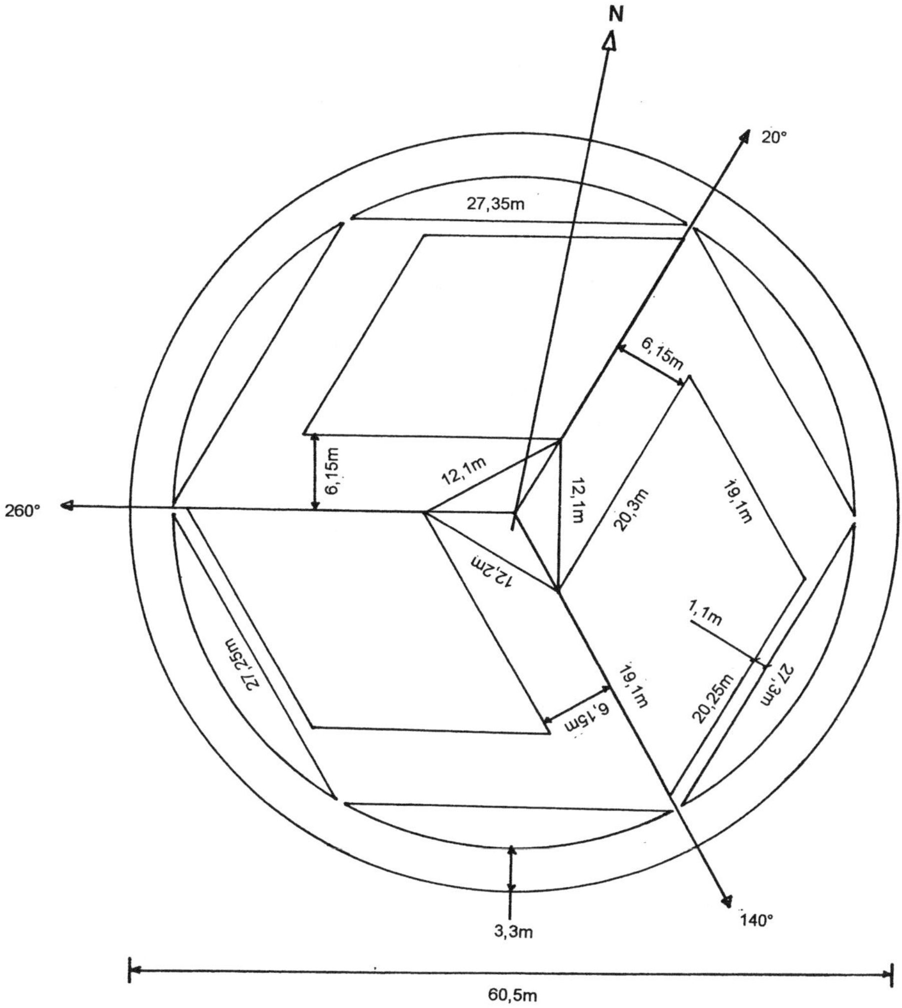
N
20°
27,35m
6,15m
260°
6,15m
12,1m
12,1m
20,3m
19,1m
12,2m
1,1m
27,25m
19,1m
20,25m
27,3m
6,15m
3,3m
140°
60,5m

Hackpen Hill, Wiltshire, England

Discovered July 4, 1999, in wheat.

Diameter: 120 meters (394 feet)

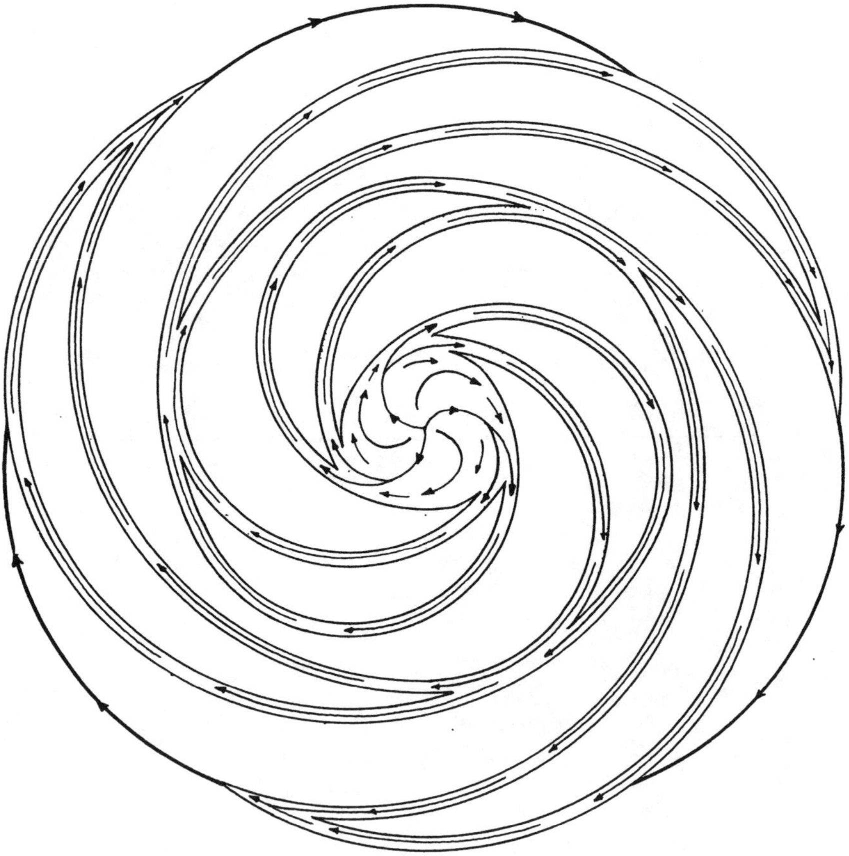

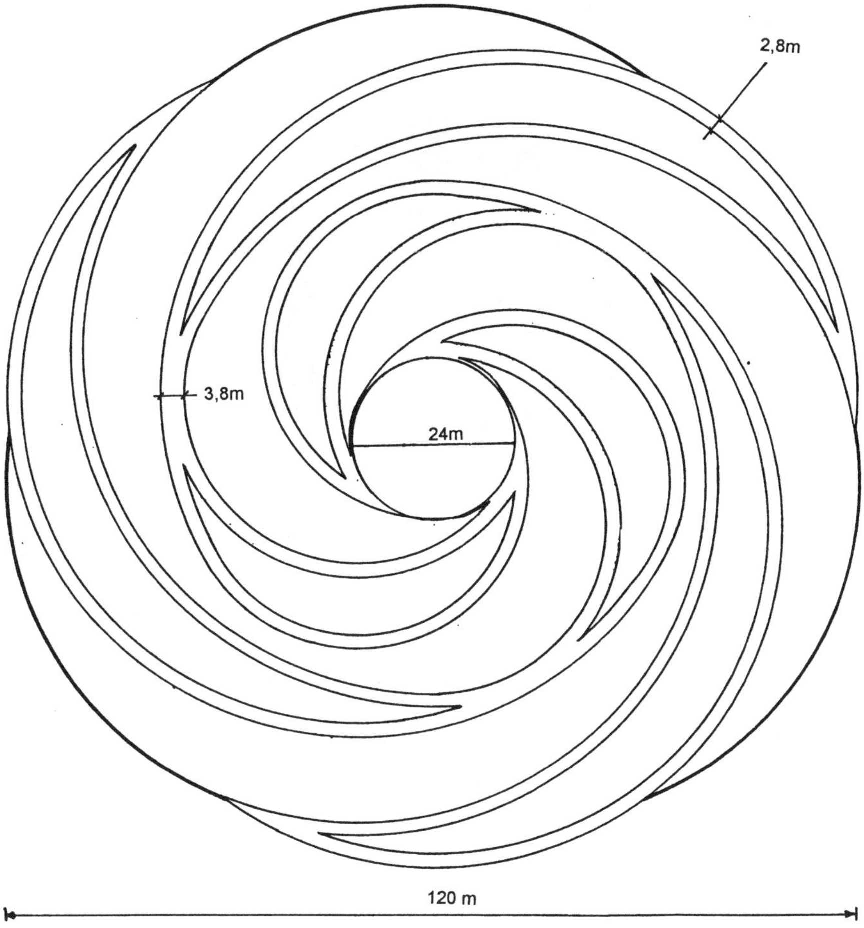
2,8m
3,8m
24m
120 m

Windmill Hill, Wiltshire, England

Discovered July 16, 1999, in wheat.

Diameter: 100 meters (328 feet)

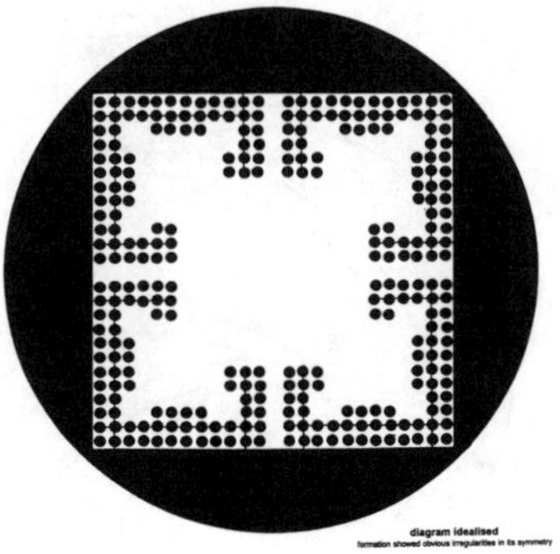

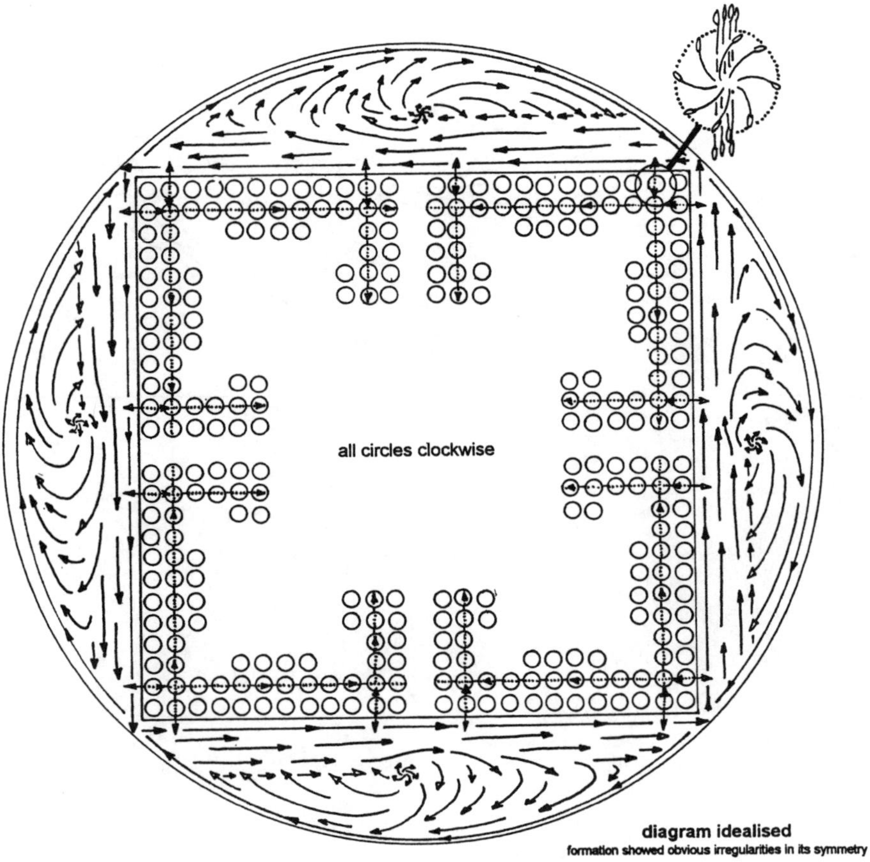

circle's size varied from 1,5 - 2,2m diameter
placing, spacing and size was irregular and showed no defined pattern
this was even the case with the paths

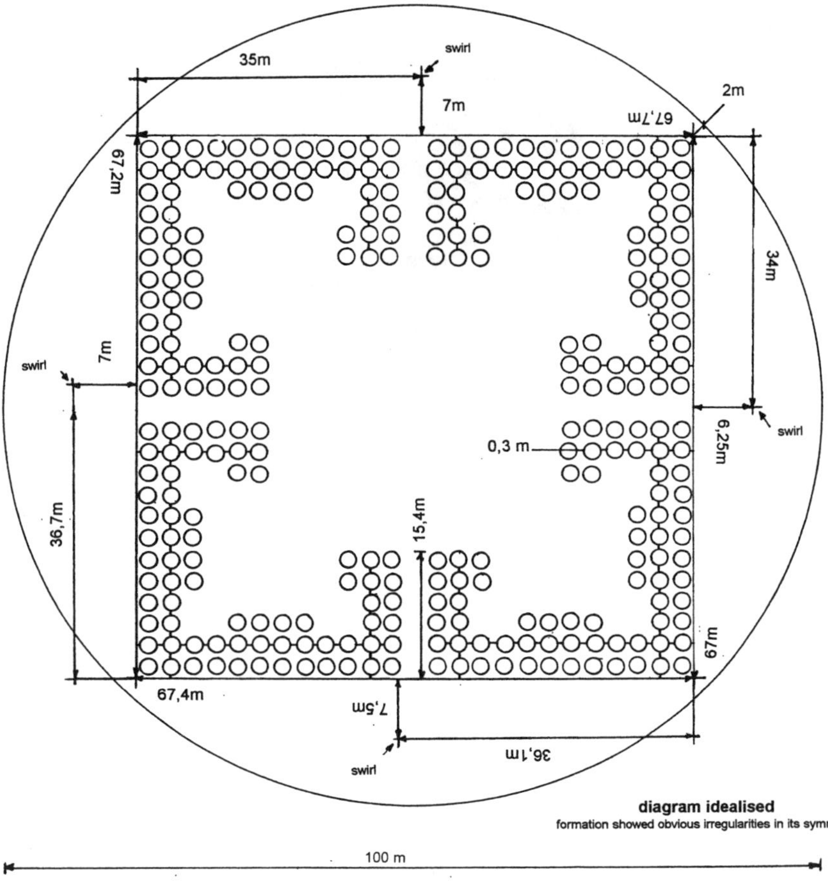

diagram idealised
formation showed obvious irregularities in its symmetry

Cherhill White Horse, Wiltshire, England

Discovered July 17, 1999, in wheat.

Diameter: 69 meters (226.38 feet)

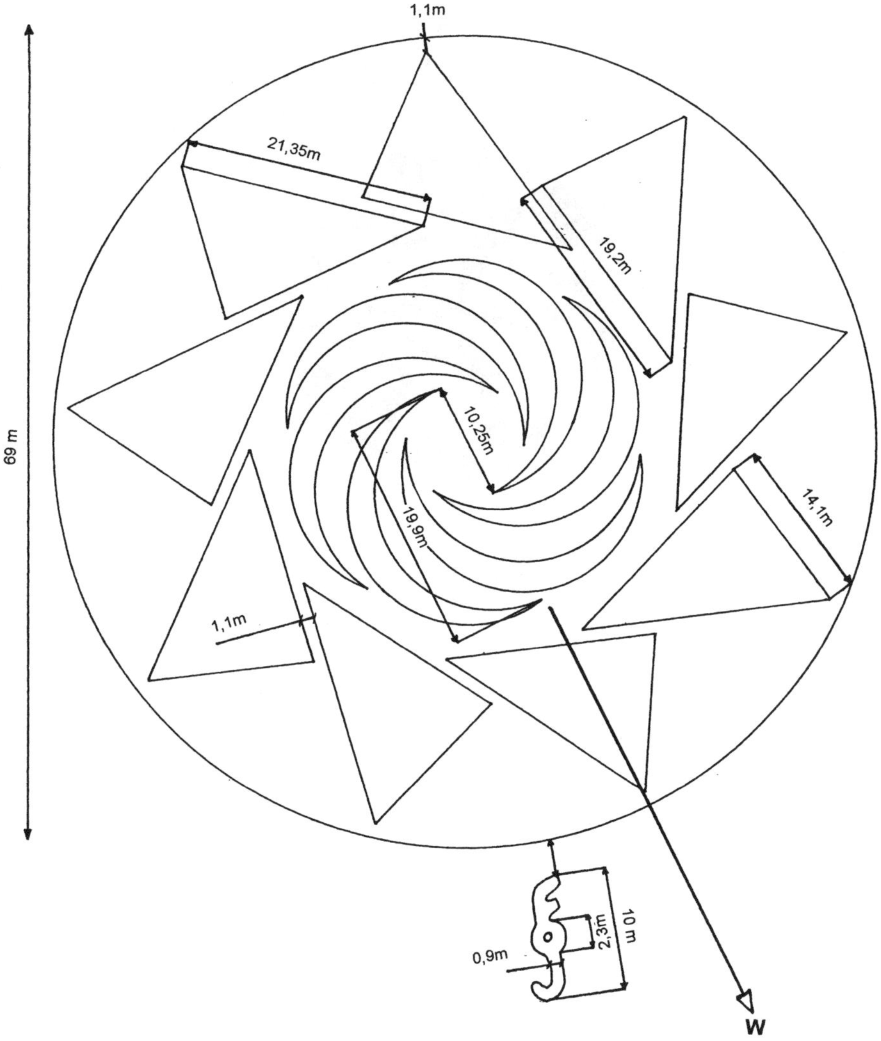
1,1m
21,35m
19,2m
69 m
10,25m
14,1m
19,9m
1,1m
10 m
2,3m
0,9m
W

Devil's Den, Wiltshire, England

Discovered July 19, 1999, in wheat.

Diameter: 61 meters (200 feet)

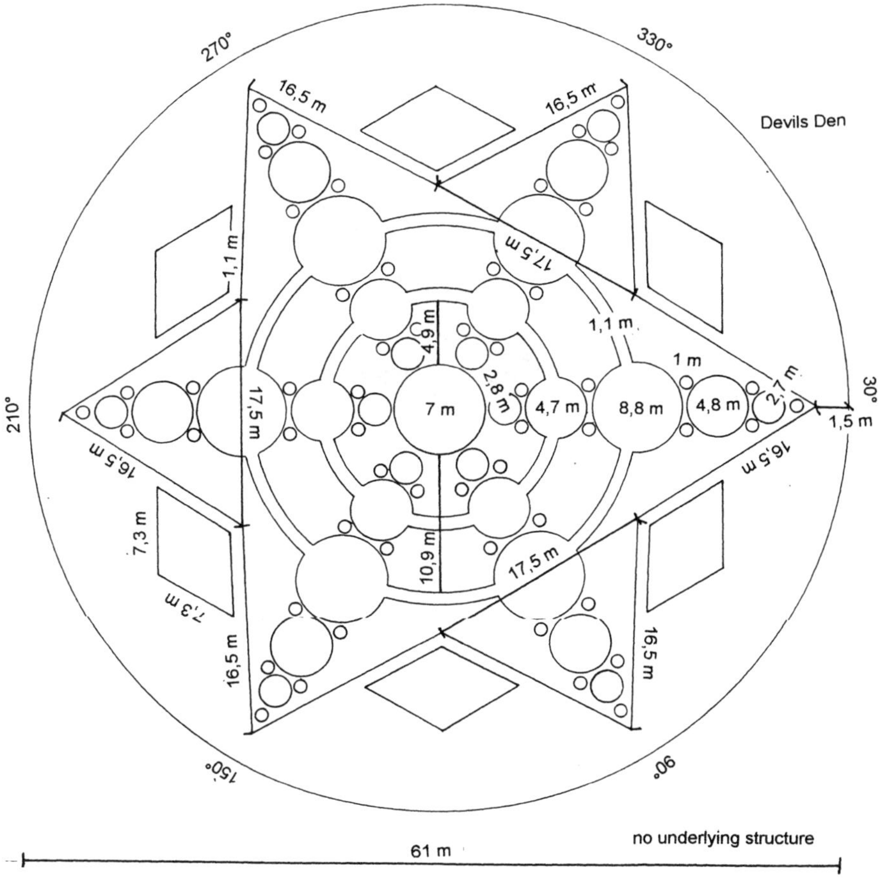

Devils Den
270°
330°
210°
30°
150°
90°
16,5 m
16,5 m
17,5 m
1,1 m
1,1 m
4,9 m
2,8 m
7 m
4,7 m
8,8 m
1 m
4,8 m
2,7 m
1,5 m
17,5 m
16,5 m
16,5 m
7,3 m
7,3 m
10,9 m
17,5 m
16,5 m
16,5 m
no underlying structure
61 m

Liddington Castle, Wiltshire, England

Discovered July 21, 1999, in wheat.

Diameter: 63 meters (207 feet)

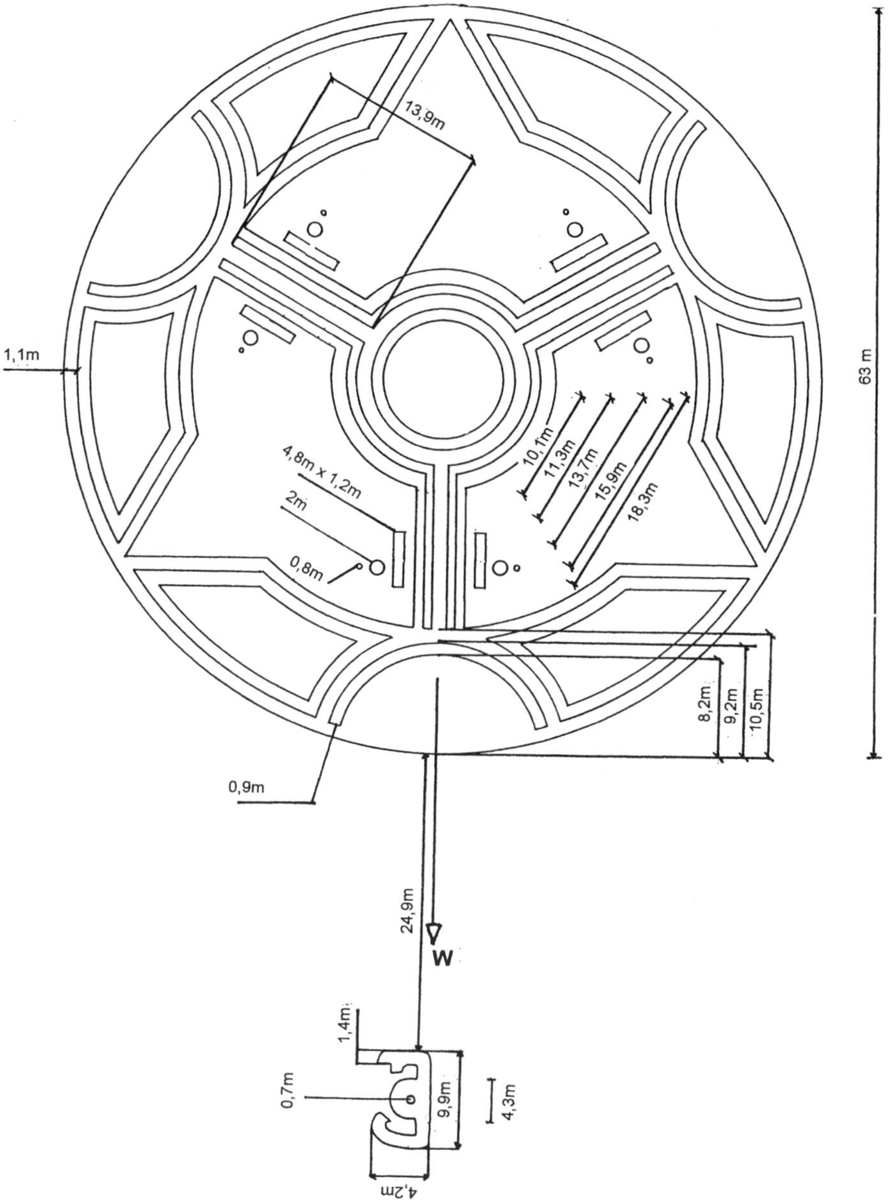
13,9m
1,1m
63 m
10,1m
11,3m
13,7m
15,9m
18,3m
4,8m x 1,2m
2m
0,8m
8,2m
9,2m
10,5m
0,9m
24,9m
W
1,4m
0,7m
9,9m
4,3m
4,2m

Barbury Castle, Wiltshire, England

Discovered July 23, 1999, in wheat.

Diameter: 79.7 meters (262 feet)

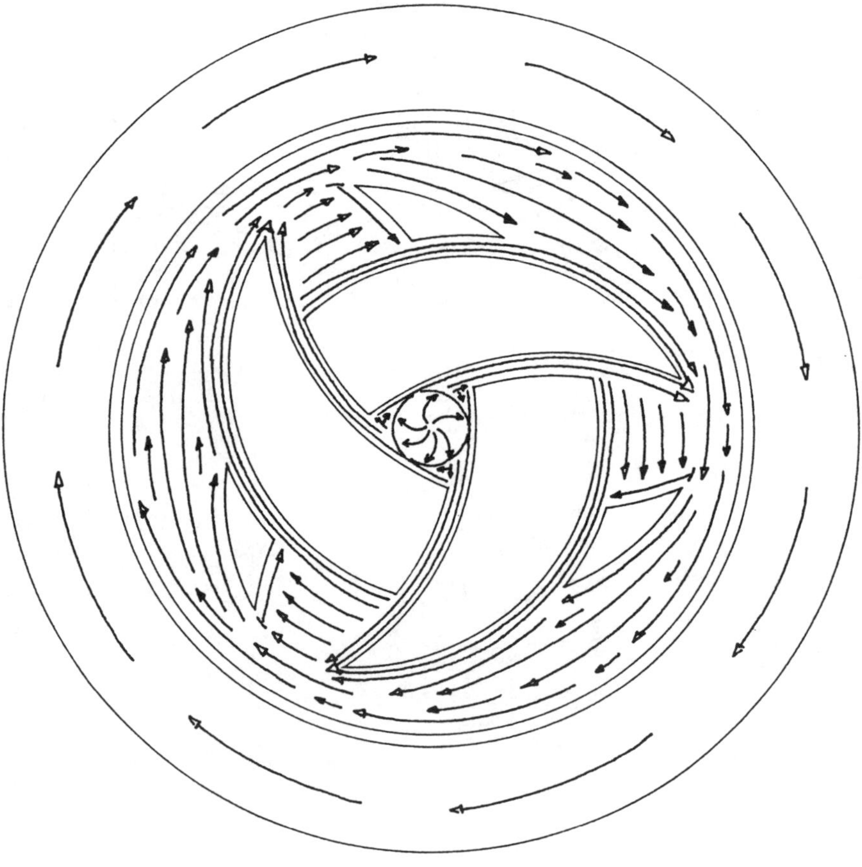

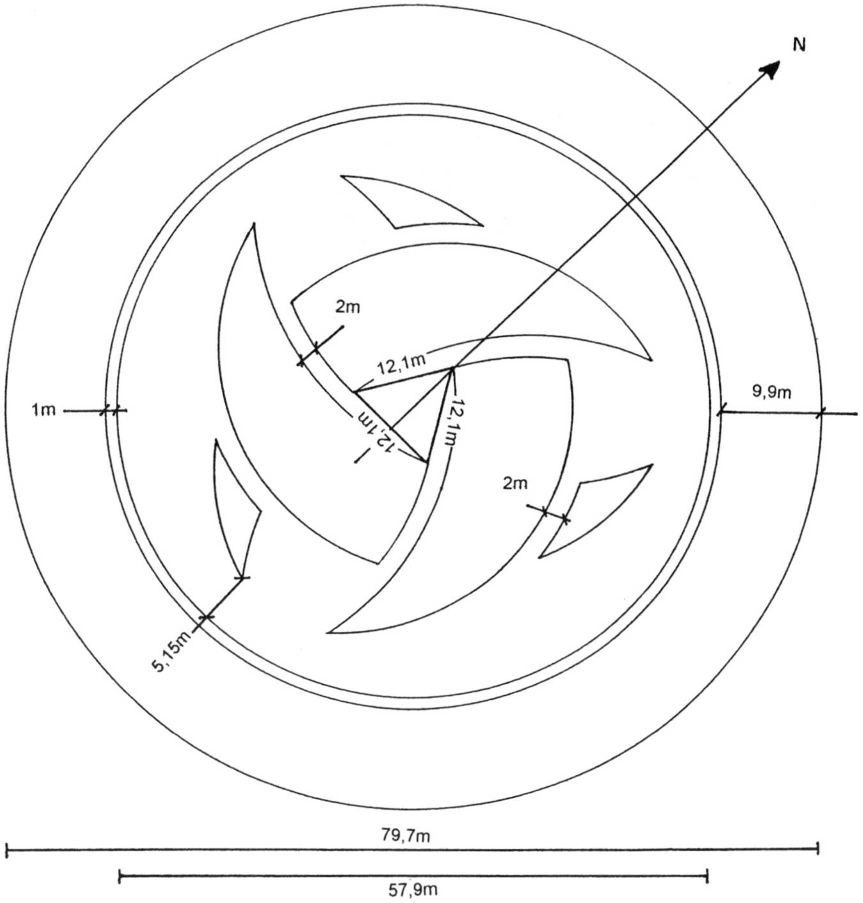
N
2m
12,1m
12,1m
12,1m
1m
9,9m
2m
5,15m
79,7m
57,9m

Silbury Hill, Wiltshire, England

Discovered July 23, 1999, in wheat.

Diameter: 97.5 meters (320 feet)

top lay

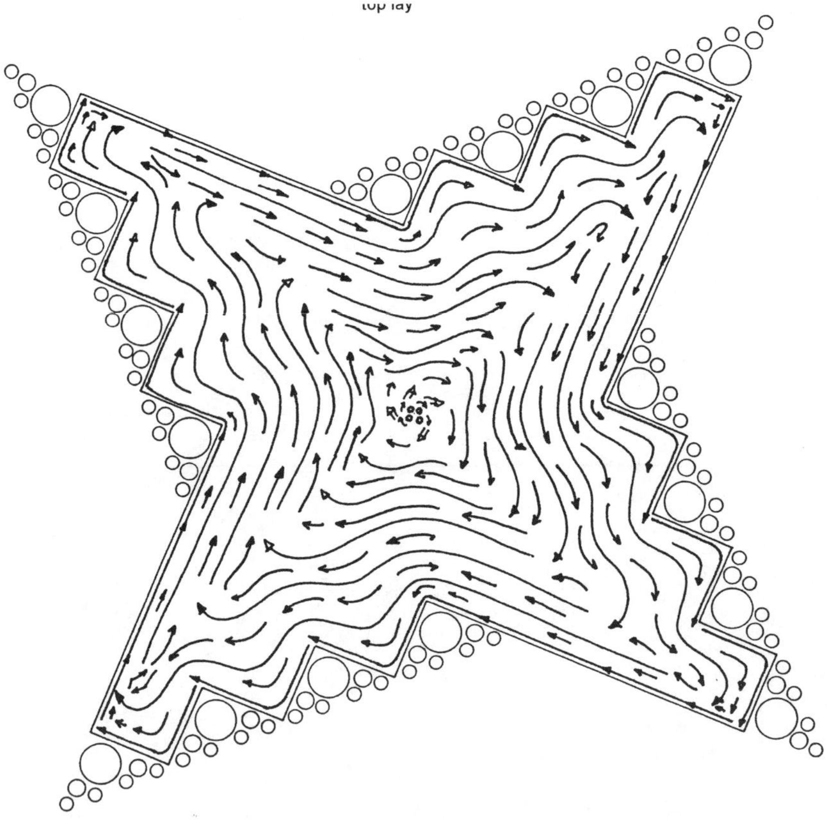

all circles were laid clockwise
some examples shown:

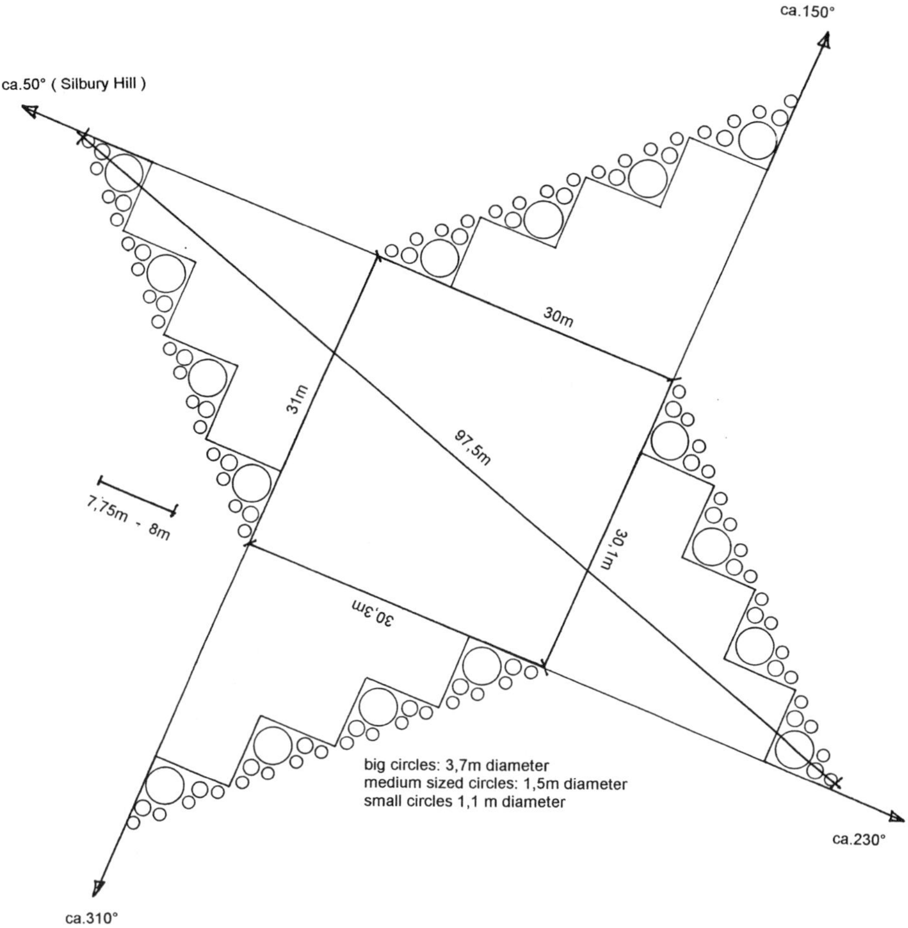
ca.150°
ca.50° (Silbury Hill)
30m
31m
97,5m
7,75m - 8m
30,1m
30,3m
big circles: 3,7m diameter
medium sized circles: 1,5m diameter
small circles 1,1 m diameter
ca.230°
ca.310°

Beckhampton, Wiltshire, England

Discovered July 28, 1999, in oats.
Diameter: 72 meters (236 feet)

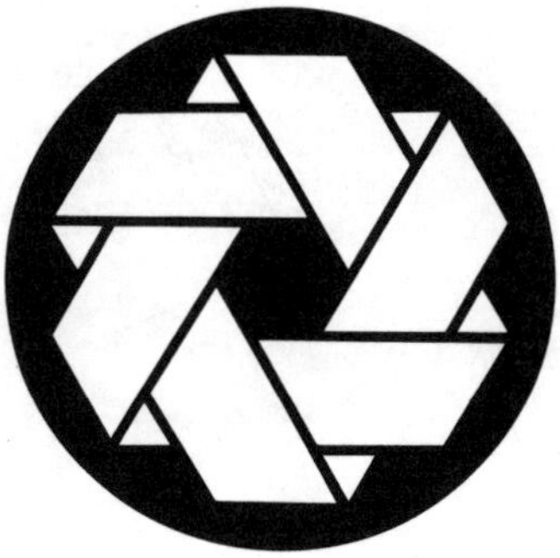

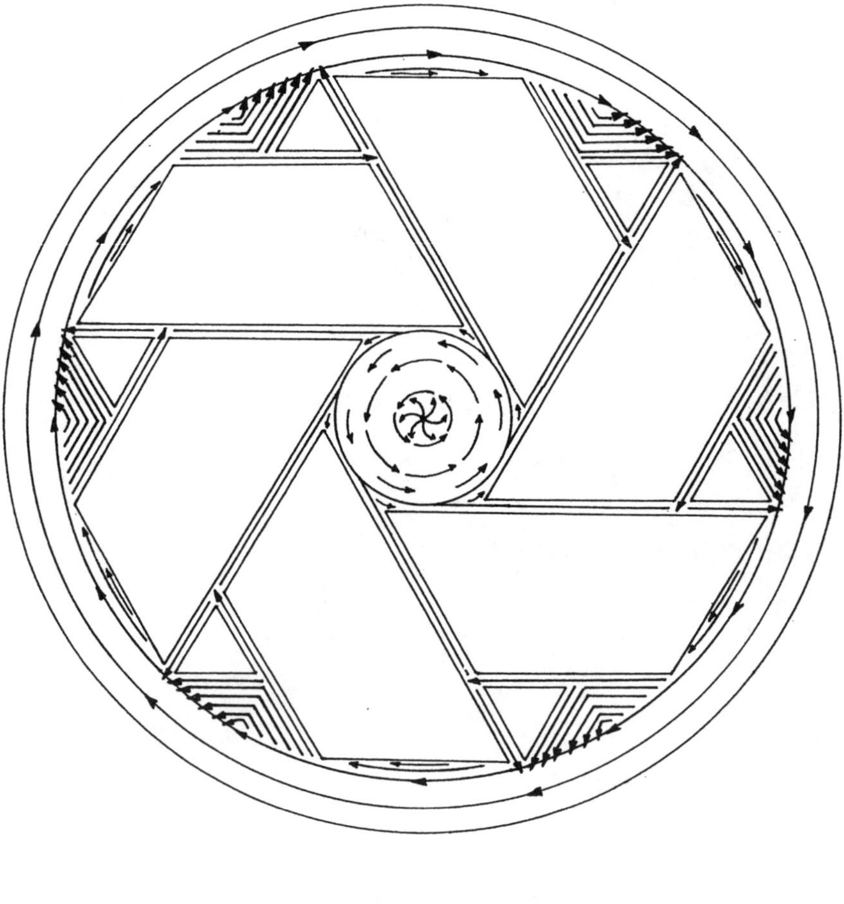

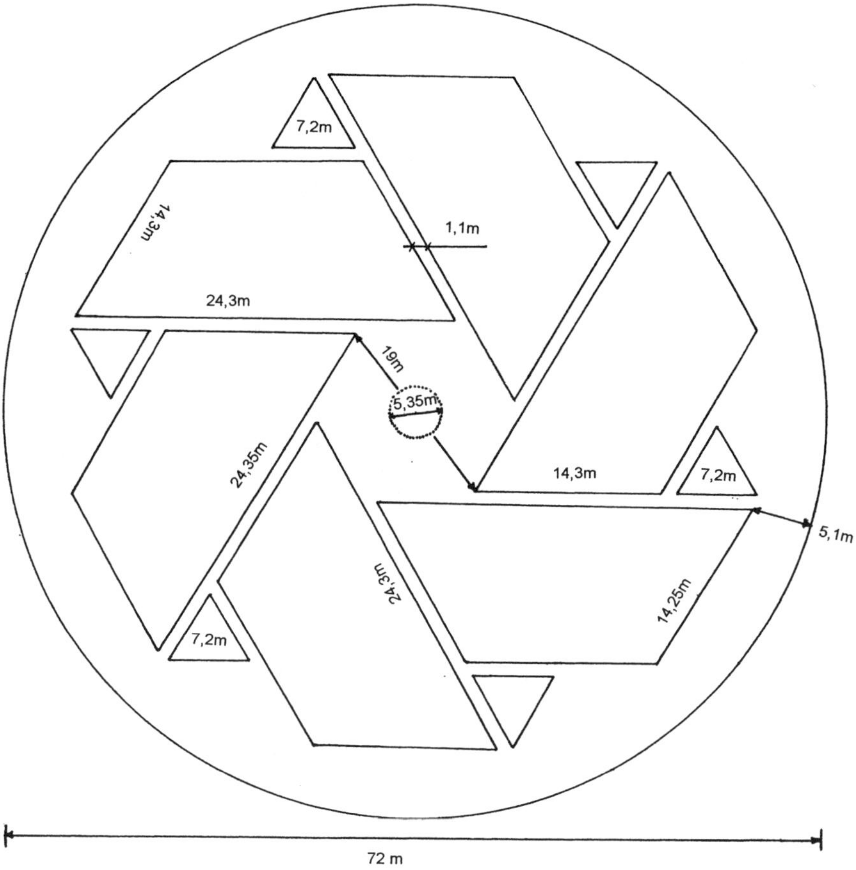
7,2m
14,3m
1,1m
24,3m
19m
5,35m
24,35m
14,3m
7,2m
5,1m
24,3m
14,25m
7,2m
72 m

Avebury Avenue, Wiltshire, England

Discovered July 31, 1999, in oats.

Diameter: 19.2 meters (63 feet)

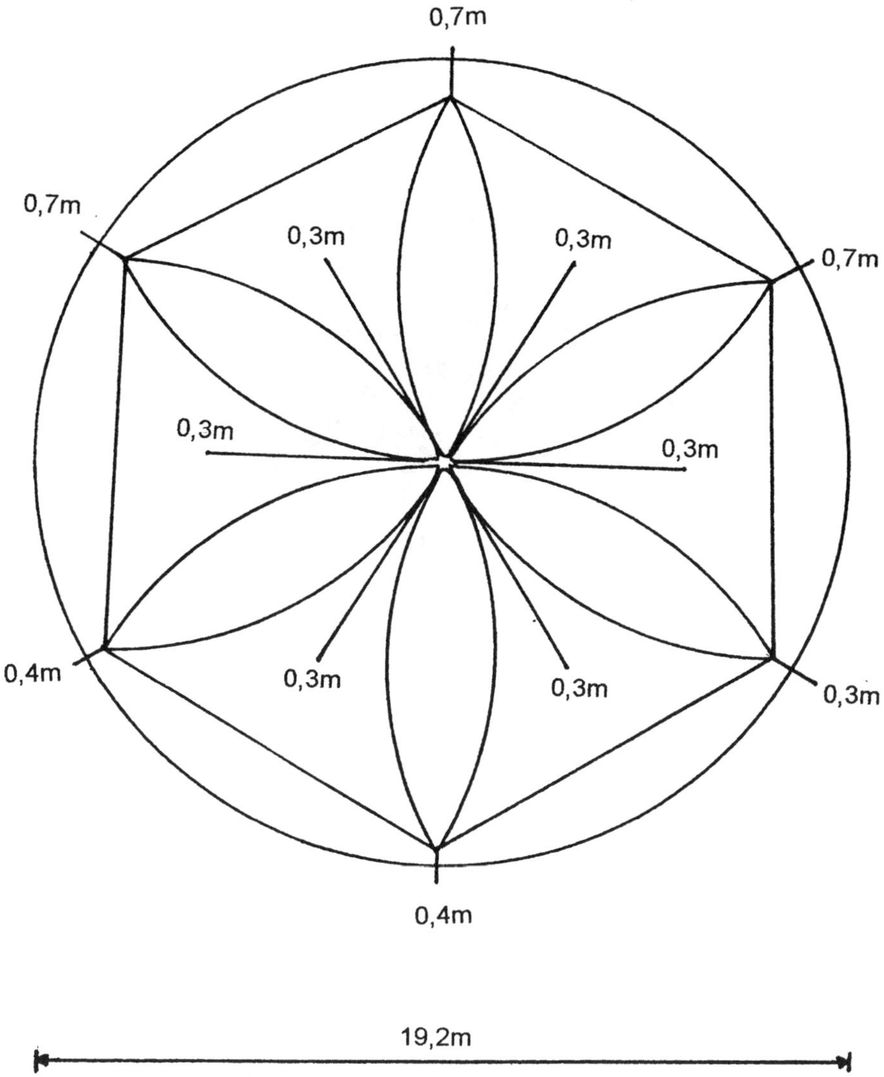
0,7m
0,7m
0,3m
0,3m
0,7m
0,3m
0,3m
0,4m
0,3m
0,3m
0,3m
0,4m
19,2m

Roundway, Wiltshire, England

Discovered July 31, 1999, in wheat.

Diameter: 110 meters (361 feet)

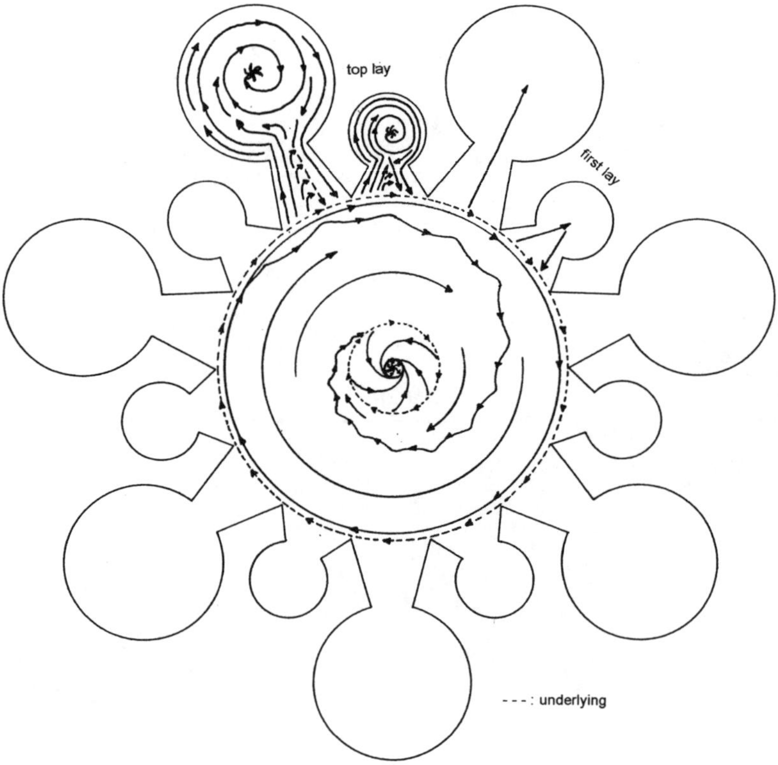

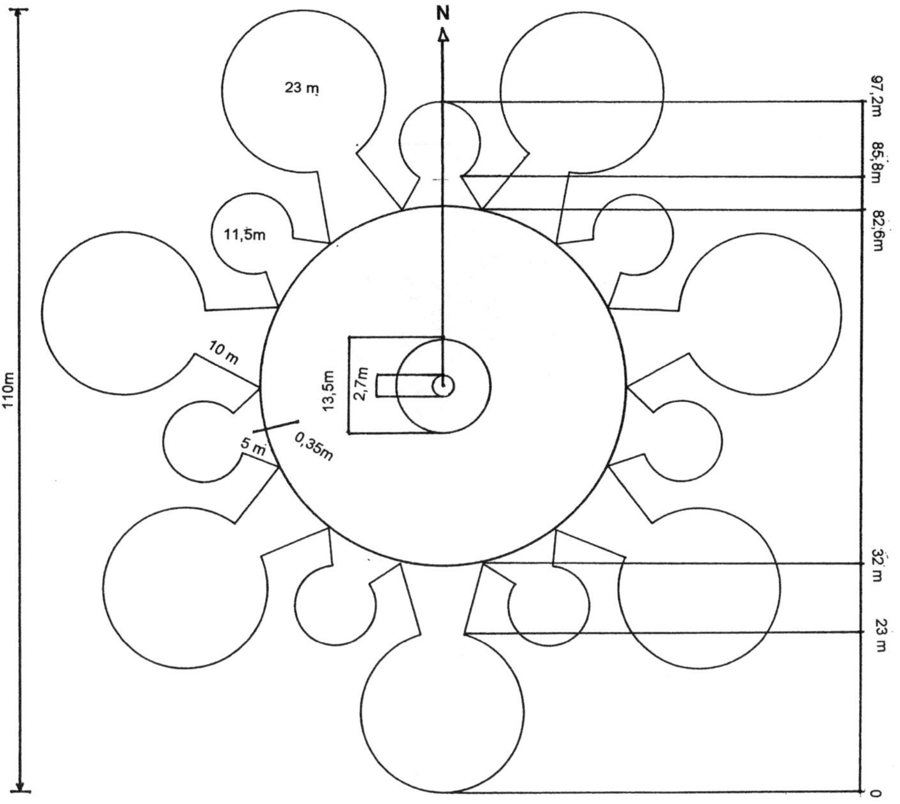
N
23 m
11,5m
10 m
5 m
0,35m
13,5m
2,7m
110m
97,2m
85,8m
82,6m
32 m
23 m
0

West Kennett Long Barrow, Wiltshire, England

Discovered August 4, 1999, in wheat.

Diameter: 100 meters (328 feet)

Top Lay

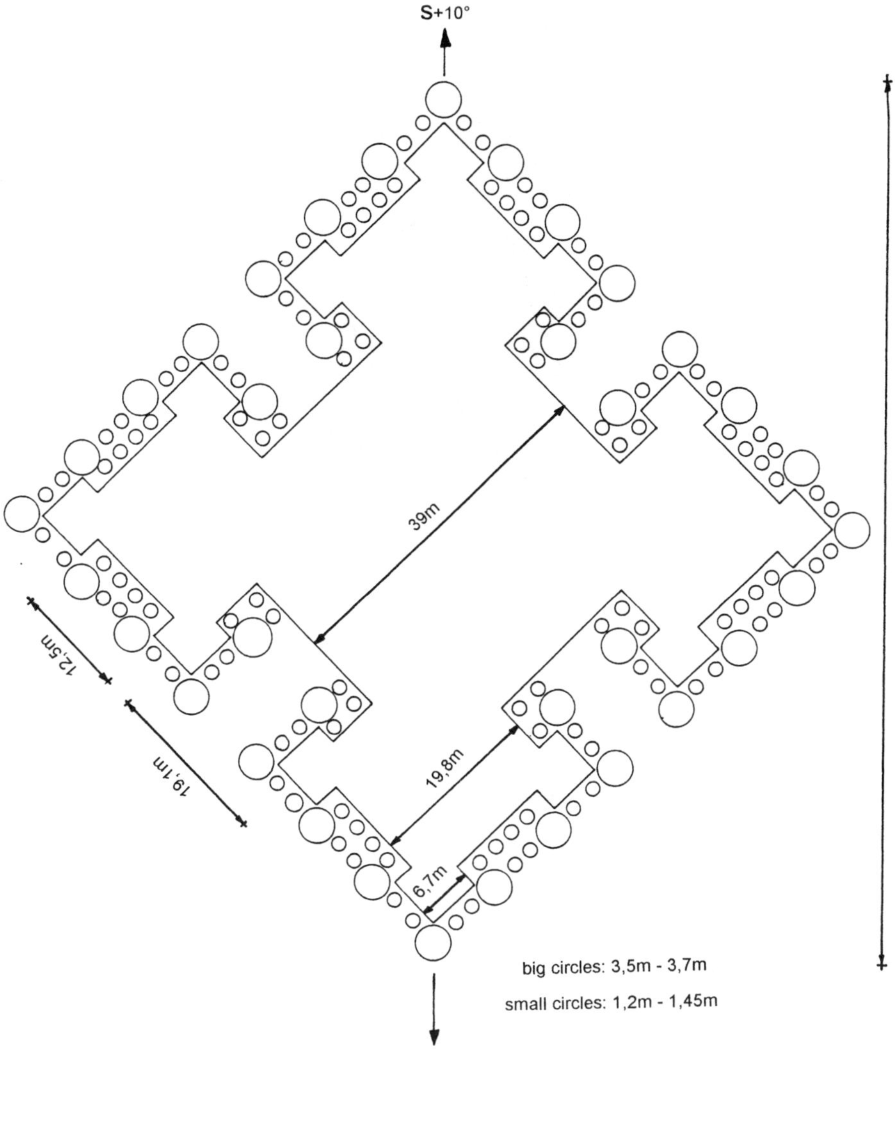
S+10°
39m
12,5m
19,1m
19,8m
6,7m
big circles: 3,5m - 3,7m
small circles: 1,2m - 1,45m

Bishops Cannings, Wiltshire, England

Discovered August 6, 1999, in wheat.

Diameter: 46.3 meters (154 feet)

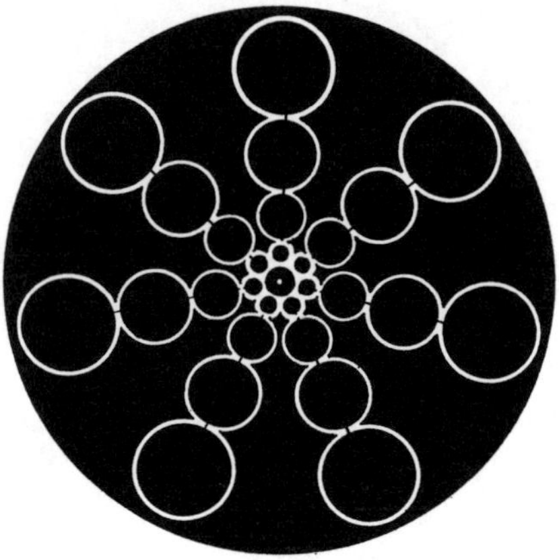

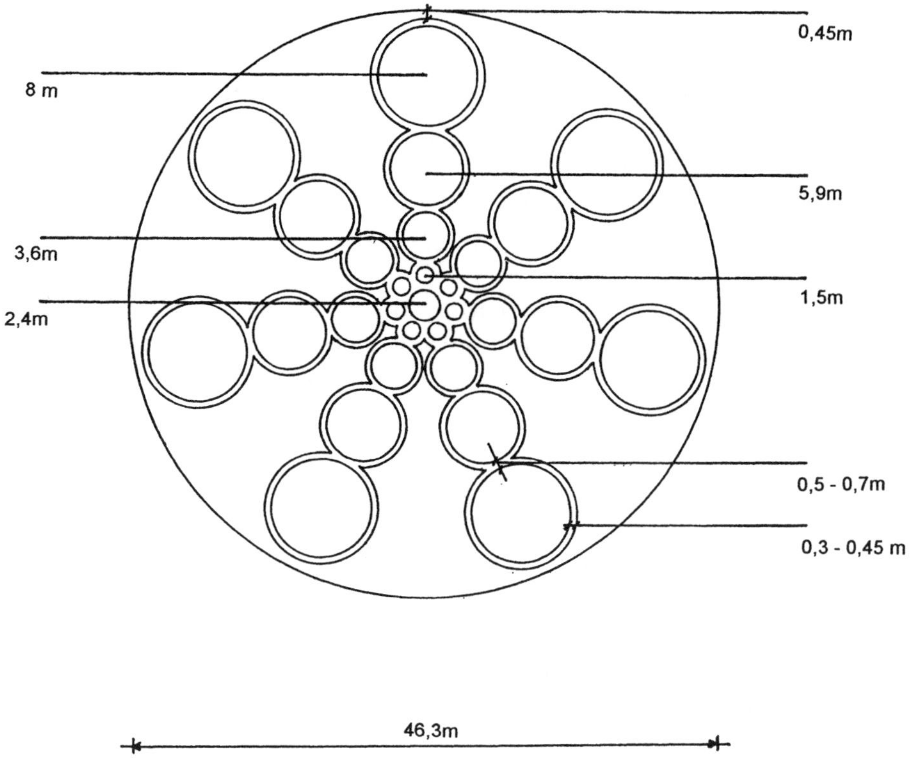
0,45m
8 m
5,9m
3,6m
1,5m
2,4m
0,5 - 0,7m
0,3 - 0,45 m
46,3m

Acknowledgments

My thanks and respect to biophysicist W. C. Levengood, Ph.D., who has had the courage to investigate and report empirical evidence about a complex phenomenon; to astronomer and mathematician Gerald Hawkins, who tried to prove it was all a hoax and could not; to attorney Michael Pill, who always asks hard questions and provides the right book and wise insights at the right time.

To John Anthony West for reading the manuscript fresh back from Australia and being provoked enough to do a foreword; to Andy Thomas for his keen and rapid editor's eye and mind in the last hours of deadlines; to Andreas Mueller, whose curiosity about the intelligence behind the formations provoked him to do surveys of extraordinary detail and elegance; to John and Julie Wakefield for their provocative study of Wiltshire's "legendary landscapes."

My deep thanks to many others for their help who are also trying to understand and photograph: ilyes who long ago sensed non-human energies were behind the global formations and invited me to share a special day at Stonehenge; Bob and Gil Nicholas and Gordon Stewart for their friendship, adventurous spirits and teaching me how to do pole shots; Lucy Pringle, who has worked hard to document the long history of circles in fields and eyewitnesses who have experienced energies not understood; Nancy Talbott, who has sustained plant and soil samplings for Dr. Levengood with professional vigor; Bert Janssen and Janet Ossebaard of The Netherlands, who helped so much with photographs, slides, video frames, surveys and interviews; Peter Sorensen, who has videotaped a valuable archive of Wiltshire crop formations since the early 1990s; Foeke Kootje and Conny de Bruyn from The Netherlands; England's Steve Alexander, Colin Andrews, Francine Blake, Donald Fletcher, Timothy Good, Michael Green, John Michell, Busty Taylor, Paul Vigay, and George Wingfield; Germany's Ulrich Kox, Hans-Juergen Kyborg, Joachim Koch, M.D., Wolfgang Schindler, Constantin and Dominic Von Durkheim; Canada's Paul Anderson, Chad and Gwen Deetken; America's Tom Burdic, John Burke, Simeon Hine, Eric Liden, Larry G. Thomas, Melodie Watts, and Michael Yudowitz.

A special thanks to Charles and Frances Mallett for their generous time, hospitality, interviews, drawings and two radio programs in the wee hours of the night from their home in Wiltshire; to Nick Nicholson for being there all the way with photographs and fresh field information, and to his colleagues Carl Nevin and Mark Haywood for sharing their video frames; to Frank Laumen for helping with all his good photographs and spring 2000 updates on the many formations in Germany; to Andy and Vittoria Abercrombie and Derick Rawson for their wonderful friendship, technical assistance and that amazing cell phone/lap top connection in the Reims, France adventure; to Kris and Ed Sherwood for their generous time, photographs, videotapes and persistence in exploring the visible and nonvisible luminosity phenomenon; to Alan Marsh and Jonathan Williams for showing me Roundway through their eyes and photographs; to Werner Anderhub from Switzerland for sharing his encounter experience and drawings about the clouds of light in the 1994 Galaxy formation; to Gary Anthony, Jim Carten, Brian Grist, Michael Hesemann and The Philosophical Research Society in Los Angeles for contributing insights from Mystery School and alchemical literature and symbols; to Lara Stefancin-Nelson, who shared the Liddington adventure, her translucent spheres and other light anomaly photographs; to Nikki Saville, who described the lights she and her brother watched for hours and their discovery of a huge crop formation where the lights had been; to Bruce Ratcliff, who persists after facts about crop formations and microwave pollution; to the keepers of The Barge Inn for sustaining the crop circle news center that the back room has become; and to Terry and June Weeks and Sharon Perry for their kindness at The Golden Swan.

The physical production of this book was enabled by Ann Douden, a superb graphic artist with a fine book sense. She has designed the formats and covers for each of my previous books and has my heartfelt gratitude. Tim Bauer, a great sculpture designer and graphic artist, drew each map by hand. Lori Kranz did a thorough proofreading and Adam Scott and Beth Kotzian were very helpful with manuscript assemblies. Jostens Press and Werner Riefling of Paper Chase Press brought it all into physical form.

Contacts for More Information

Steve Alexander, Photographer
Karen Douglas, Writer
Temporary Temple Press
27 St. Francis Road
Gosport, Hampshire
England PO12 2UG
Tel/Fax: (44) 023-9235-2867
e-mail: temporarytemples@netscapeonline.co.uk
Website: http://www.cropcircleconnector.com/temporarytemples/homepage.html

Werner Anderhub, Photographer and Author
Das Geheimnis der Kornkreise (Book: *The Mystery of Crop Circles* © 2000)
Schillingstrasse 28
CH-3005 Bern
Switzerland
Tel/Fax (41) 31-352 3641
e-mail: werner@kornkreise.ch
Website: http://kornkreise.ch

Paul Anderson, Director
The Millennium Project
2086 West 2nd Avenue, Suite 202
Vancouver, British Columbia
Canada V6J 1J4
Tel/Fax: 604-731-8522
Cell: 604-727-1454
e-mail: psa@direct.ca
Website: http://persweb.direct.ca/psa

Colin Andrews, Author
Circles Phenomena Research International
P. O. Box 3378
Branford, CT 06405-1978
Tel/Fax: 203-483-0822
e-mail: CPRAndrews@worldnet.att.net
Websites: http://home.clara.net/lovely/cprhome.html
http://cropcircleconnector.com/CPII/cprl.html

Francine Blake
Wiltshire Crop Circle Study Group (WCCSG)
P. O. Box 939
Devizes, Wiltshire
England SN10 3TA
e-mail: sky.picture@virgin.net

John Burke
Electrostatic Research
20 Cyrus Point Road
Bayville, New York 11709
Fax: 516-628-0447
e-mail: ProSeedTec@aol.com

Ian Christopher, Filmmaker
Mariposa-Pacific Research Institute
Something Wonderful Has Happened But It Was Not in the News! (1999 documentary)
Palo Alto, California
Tel: 650-266-9236
e-mail: mariposapacific@yahoo.com
Film orders: http://www.eastwest.com

Mark Fussell and
Stuart Dike, Editors
http://www.cropcircleconnector.com
e-mail: mjfussell@marque.demon.co.uk

Simeon Hein, Ph.D.
Mount Baldy Institute for Resonant Viewing
1942 Broadway, Suite 409
Boulder, CO 80302
Tel: 303-440-7393
e-mail: Simeon@mountbaldy.com
Website: www.mountbaldy.com

Michael Hesemann, Editor
Magazin 2000
Warringer Strasse 1
D-4000 Dusseldorf 1
Germany
Fax: (49)-211-354-893
e-mail: webmaster@m-n-d.com
Website: http://hesemann.m-n-d.com

Bert Janssen, Videographer, Writer and Producer
Crop Circles - The Research (1999 documentary)
P. O. Box 10
9460 AA Gieten
The Netherlands
Tel: (31)-05-98-468023
e-mail: info@bertjanssen.nl
Website: http://www.bertjanssen.nl

Foeke Kootje, Videographer
Conny de Bruyn
UFOCUS
Hopperstrasse 9
8602 AP Sneek
The Netherlands
Tel: (31)-0515-421-473
e-mail: UFOCUS@home.nl

Frank Laumen, Photographer
Crop Circle Connections: Intangible Realms of Reality (2000 documentary)
Dahlemerstrasse 12
51377 Leverkusen
Germany
Tel: (49) 214-909-8070
e-mail: laumen@invisiblecircle.de
Websites: http://www.franklaumen.de/Anfang.htm
http://www.invisiblecircle.de

Andreas Mueller
I.C.C.A. — International Crop Circle Archive and
(ic) invisiblecircle
Crop Circle Formations — Diagram Trilogies 1998-2000 available from
http://invisiblecircle.de
http://invisiblecircle.de Shop
Johannisstrasse 5
66111 Saarbruecken
Germany
Tel: (49) 681-399 945
e-mail: mueller@invisiblecircle.de
Website: http://invisiblecircle.de

Charles and Frances Mallett, Photographer and Artist
C&F Phenomenon Research
e-mail: cfphenomenonresearch@hotmail.com
Website: http://www.cropcircleconnector.com/signsofthetimes/signsofthetimes.html

Nick Nicholson, Editor
The Circular Review
6D Pond View
Moor Farm, Moore Lane
Calverton, Nottinghamshire
England NG14 6FZ
Tel: (44) 0115-965-3488
Cell: (44) 097-6212087
e-mail: circular99@yahoo.com
Website: http://cropcircleconnector.com/column/review.html

Janet Ossebaard, Photographer and Author
Graancirkels. Een wereldwyd mysterie. (Book: *Crop Circles: A Worldwide Mystery* © 2000)
P. O. Box 10
9460 AA Gieten
The Netherlands
Tel: (31) 05-98-468023
e-mail: ossebaard@wxs.nl

Philosophical Research Society, Publisher
The Secret Teachings of All Ages © 1928 by Manly Palmer Hall
3910 Los Feliz Boulevard
Los Angeles, California 90027-2399
Tel: 323-663-2167
Fax: 323-663-9443

Lucy Pringle, Photographer and Author
5 Town Lane
Sheet
Petersfield
Hampshire
England GU32 2AF
Tel/Fax: (44) 01730-263454
e-mail: LucyPringle@cs.com
Website: http://home.clara.net/lucypringle

Ron Russell, Photographer and Artist
Founder, Midwest Research Group
Center for Crop Circle Studies
P. O. Box 460760
Aurora, Colorado 80046
Tel: 303-400-1322
e-mail: ronr@street.com
Website: http://www.cropcircles.org

Ed and Kris Sherwood
Millennium Research
P. O. Box 2084
Santa Monica, CA 90406
Tel: 310-319-9329
e-mail: millennium9@earthlink.net
Website: http://cropcircleconnector.com/Millennium/millennium.html

Kallista and Chet Snow, Ph.D.
Sedona Conferences
P. O. Box 1738
Sedona, Arizaon 86339
Fax: 520-204-1955
e-mail: chetsnow@sedona.net
Website: http://people.delphi.com/sedonaconf

Peter Sorensen, Videographer
c/o 1115 East Cordova #219
Pasadena, CA 91106
Tel: 626-793-8010
e-mail: croppie2@yahoo.com
Website: http://cropcircleconnector.com/Sorensen/PeterSorensen99.html

Nancy Talbott, Coordinator
BLT Research
P. O. Box 400127
Cambridge, Mass. 02140
Tel: 617-492-0415
Fax: 617-492-0414
e-mail: bltresearch@mediaone.net
Website: http://www.bltresearch.net

Busty Taylor, Photographer
52 Appletree Grove
Andover, Hampshire
England SP10 3RG
Tel: (44) 01264-324-496
Website: http://www.aviation-uk.com/cropper/98busty2.htm

Andy Thomas, Editor
SC: The Bimonthly Journal of Crop Circles and Beyond
36 Graham Crescent
Mile Oak
Portslade, East Sussek
England BN41 2YB
Tel: (44) 01273-885-117
e-mail: scr@landl.freeserve.co.uk
Website: http://www.cropcircleconnector.com/SC/SC.html

John and Julie Wakefield, Authors
Nod Press
P. O. Box 338
Marlborough, Wiltshire
England SN8 4PW
Tel/Fax: (44) 01672-851140
e-mail: nod-press@merlinwebservices.co.uk
Website: http://www.merlinwebservices.co.uk/nod-press

John Anthony West, Egyptologist, Author and Egypt Tours
4 Second Street
Athens, New York 12815
Fax: 518-945-2329
e-mail: jawsphinx@aol.com

Bibliography

ALEXANDER, Steve and Karen Douglas. *Crop Circle Year Book 1999.* Temporary Temple Press, U. K., © 1999.

ANDERHUB, Werner, and Hans Peter Roth. *Das Geheimnis der Kornkreise* (The Mystery of Crop Circles). Verlag, Aarau, Switzerland, © 2000.

BARTHOLOMEW, Alick. *Crop Circles, Harbingers of World Change.* Gateway Books, U. K., © 1991.

BOHM, David. *Wholeness and the Implicate Order.* Cox & Wyman Ltd., © 1980.

BORD, Janet and Colin. *Unexplained Mysteries of the 20th Century.* Contemporary Books, Chicago, © 1989.

CAPT, E. Raymond. *The Traditions of Glastonbury.* Artisan Sales, California © 1983.

CRAWFORD, Harriet. *Sumer and the Sumerians,* Press Syndicate, University of Cambridge, U. K., © 1991.

CRUMP, Thomas. *Solar Eclipse.* Constable and Co., Ltd., U. K., © 1999.

DAVIS, Beth, Editor. *Ciphers in the Crops, The Fractal and Geometric Circles of 1991.* Gateway Books, U. K., © 1992.

DELGADO, Pat and Colin Andrews. *Crop Circles, The Latest Evidence.* Bloomsbury Publishing Ltd., U. K., © 1990.

DOBSON, Rev. C. C. *Did Our Lord Visit Britain As They Say In Cornwall and Somerset?* Covenant Publishing, U. K., © 1936-1993.

ENCYCLOPAEDIA BRITANNICA. Encyclopaedia Britannica, Inc., © 1993 and 1972.

FONTANA, David. *The Secret Language of Symbols.* Chronicle Books, California, © 1993.

GLEICK, James. *Chaos, Making A New Science.* Penguin Books, New York, © 1987.

HALEVI, Zev ben Shimon. *Kabbalah, Tradition of Hidden Knowledge.* Thames and Hudson, U. K., © 1979.
Kabbalah, The Divine Plan. HarperCollins, New York, © 1996.

HALL, Manly Palmer. *The Secret Teachings of All Ages: An Encyclopedic Outline of Masonic, Hermetic, Qabbalistic and Rosicrucian Symbolical Philosophy.* The Philosophical Research Society, Los Angeles, California, © 1928.

HARDCASTLE, F. *The Chalice Well.* Chalice Well Trust, Glastonbury, © 1990.

HAWKINS, Gerald S. with John B. White. *Stonehenge Decoded.* William Collins, U. K., © 1965.
Mindsteps to the Cosmos. Harper & Row, New York, © 1983.
"Geometry in English Wheat Fields." *Mathematics Teacher,* Vol. 88, page 802, 1995; and Vol. 91, page 441, 1998.
"Diatonic Ratios." *Cereologist,* No. 21, 1998.
"From Euclid to Ptolemy in English Crop Circles." *Bulletin of the American Astronomical Society,* Vol. 29, No. 5, page 1263, 1997.
"Theorems in Wheat Fields." *Science News,* Vol. 150, page 239, 1996.
"Solar System." *CPRI Newsletter,* Vol. 4, No. 2, 1995.
"Oliver's Castle." *Cereologist,* No. 8, 1993.

"Music of the Spheres." *Share International*, Vol. 11, No. 10, 1992.
"Euclid's Crop Circles," and "Letter." *Science News*, Vol. 141, No. 5, pages 76 and 156, 1992.
"Math in Crop Circles." *Cereologist*, No. 6, 1992.
"Probing the Mystery of Those English Crop Circles." *COSMOS Journal of Emerging Issues*, Vol. 2, No. 1, page 23, 1992.

HESEMANN, Michael. *The Cosmic Connection.* Gateway Books, U. K., © 1996.
Botschaft aus dem Kosmos/Kornkreise (The Cosmic Connection), © 1992/1996.
Die kommende Weltkrise (The Coming World Crisis), © 1998.
Mystery of the Crop Circles, © 1991. (Video)

HOWE, Linda Moulton. *Glimpses of Other Realities, Volume I: Facts & Eyewitnesses.* LMH Productions, Philadelphia, Pennsylvania © 1995.
Glimpses of Other Realities, Volume II: High Strangeness. Paper Chase Press, New Orleans, Louisiana, © 1998.

ISRAELI, Yael. *In the Light of the Menorah, Story of a Symbol.* The Israel Museum, Jerusalem, © 1999.

JANSSEN, Bert. *Crop Circles — The Research*, © 1999.
Crop Circles: What On Earth Is Going On? © 1997. Bert Janssen and Janet Ossebaard, The Netherlands.

LAWLOR, Robert. *Sacred Geometry, Philosophy and Practice.* Thames and Hudson, © 1982.

LEET, Leonora. *The Secret Doctrine of the Kabbalah, Recovering the Key to Hebraic Sacred Science.* Inner Traditions International, Vermont, © 1999.

LEVENGOOD, W. C. and Talbott, Nancy. "Dispersion of Energies in Worldwide Crop Formations." *Physiologia Plantarum*, The Netherlands, 105: 615-624, 1999.

LEVENGOOD, W. C. and Burke, John. "Semi-Molten Meteoric Iron Associated with A Crop Formation." *Journal of Scientific Exploration*, Santa Fe, New Mexico, Vol. 9, No. 2, pp. 191-199, 1995.

LEVENGOOD, W. C. "Anatomical Anomalies in Crop Formation Plants." *Physiologia Plantarum*, The Netherlands, 92: 356-363, 1994.

LEWIS, Lionel Smithett. *St. Joseph of Arimathea At Glastonbury.* James Clarke & Co., U. K., © 1922 and 1955.

LUNDY, Miranda. *Sacred Geometry.* Wooden Books, U. K., © 1998.

MANN, Nick. *Glastonbury Tor, A Guide to the History and Legends.* Annenterprise, U. K., © 1996.

MATTHEWS, John. *A Glastonbury Reader, Selections from the Myths, Legends and Stories of Ancient Avalon.* The Aquarian Press, U. K., © 1991.

MICHELL, John. *New Light on the Ancient Mystery of Glastonbury.* Gothic Image Publications, U. K., © 1990 and 1997, and *The Cereologist*, No. 7, © 1992.

MUELLER, Andreas. *Crop Circle Formations — Diagram Trilogies '98.* A. Mueller, © 1998.
Crop Circle Formations — Diagram Trilogies '99. A. Mueller, © 1999.
Crop Circle Formations — Diagram Trilogies 2000. A. Mueller, © 2000.

NOYES, Ralph, Editor. *The Crop Circle Enigma, Grounding the Phenomenon in Science, Culture and Metaphysics.* Gateway Books, U. K., © 1990.

OSSEBAARD, Janet. *Crop Circles — The Research,* © 1999.
Crop Circles: What On Earth Is Going On? © 1997. Bert Janssen and Janet Ossebaard, The Netherlands.

PENNICK, Nigel. *Sacred Geometry, Symbolism and Purpose in Religious Structures.* Capall Bann Publishing, U. K., © 1994.

PIGGOTT, Stuart. *The Druids.* Thames and Hudson, U. K., © 1968 and 1975.

PRINGLE, Lucy. *Crop Circles, The Greatest Mystery of Modern Times.* Thorsons, U. K., © 1999.

READE, Julian. *Mesopotamia.* Trustees of the British Museum, U. K., © 1991.

SCHUHMACHER, Stephan and Gert Woerner, editors. *The Encyclopedia of Eastern Philosophy and Religion: Buddhism, Hinduism, Taoism and Zen.* Shambhala Publications, Boston, © 1994.

SCHWALLER De LUBICZ, R. A. *The Temple of Man, Volumes I and II,* © 1957. English translation from the original French by Inner Traditions International, Vermont, © 1998.

STEIN, Diane. *Essential Reiki: A Complete Guide to an Ancient Healing Art.* The Crossing Press, California, © 1995.

TAYLOR, Busty. *Crop Circles of 1991.* Beckhampton Books, U. K., © 1992.

THOMAS, Andy. *Vital Signs, A Complete Guide to the Crop Circle Mystery and Why It Is NOT A Hoax.* SB Publications, U. K., © 1998.
Fields of Mystery, The Crop Circle Phenomenon in Sussex. SB Publications, © 1996.
Quest for Contact, A True Story of Crop Circles, Psychics and UFOs. SB Pub., © 1997.

TOMPKINS, Peter. *Secrets of the Great Pyramid.* Harper & Row, N. Y., © 1971.

WAKEFIELD, Julie and John. *Legendary Landscapes, Secrets of Ancient Wiltshire Revealed.* Nod Press, U. K., © 1999.

WEATHERHILL, Craig and Paul Devereux. *Myths and Legends of Cornwall.* Sigma Leisure, U. K., © 1994 and 1998.

WEST, John Anthony. *Serpent in the Sky.* Quest Books, U. K., © 1993.

WESTWOOD, Jennifer, Editor. *The Atlas of Mysterious Places.* Weidenfeld & Nicolson, N. Y., © 1987.

WHITE, Ralph. *The Rosicrucian Enlightenment Revisited.* Lindisfarne Books, N. Y., © 1999.

WOOLLEY, C. Leonard. *The Sumerians.* W. W. Norton and Co., N. Y., © 1965.

Index

E

F

L

M

R

S

About the Author

Investigative reporter and author Linda Moulton Howe in Stanton St. Bernard, Wiltshire, formation on August 15, 1999.

Linda graduated from Stanford University with a master's degree in communication and has devoted her documentary film, television, writing and radio career to productions concerning science, medicine and the environment. She has received local, national and international awards, including three regional Emmys, and a national Emmy nomination; was an honored medical producer in Boston's WCVB Station Excellence Peabody Award; and received the Aviation & Space Writers Association Award for Writing Excellence in Television, the Colorado Florence Sabin Award for "outstanding contributions to public health," and a Chicago Film Festival Documentary Golden Plaque.

She was director of international programming for *Earthbeat*, an environmental series broadcast on Turner's WTBS Superstation, Atlanta; produced, directed, wrote and edited for *The World of Chemistry* (PBS) and for UNICEF about international child survival efforts; and was creator and supervising producer of *UFO Report: Sightings* (FOX). She has written four books, reports news for the Premiere Radio Networks, and is reporter and editor of www.earthfiles.com, a news website about science, medicine, the environment, and real X-files.

Linda Moulton Howe
P. O. Box 300
Jamison, PA 18929-0300
Fax: 215-491-9842
e-mail: earthfiles@earthfiles.com
Website: http://www.earthfiles.com